D0105046

Reaching for the Moon

Reaching for the Moon

A Short History of the Space Race

Roger D. Launius

Yale UNIVERSITY PRESS
New Haven & London

Published with assistance from the Mary Cady Tew Memorial Fund.

Yale University Press books may be purchased in quantity for educational, business, or promotional use. For information, please e-mail sales.press@yale.edu (U.S. office) or sales@yaleup.co.uk (U.K. office).

Set in Janson Roman type by Integrated Publishing Solutions,
Grand Rapids, Michigan.
Printed in the United States of America.

Library of Congress Control Number: 2018950632
ISBN 978-0-300-23046-8 (hardcover : alk. paper)

A catalogue record for this book is available from the British Library.

This paper meets the requirements of ANSI/NISO Z39.48-1992
(Permanence of Paper).

10 9 8 7 6 5 4 3 2 1

Contents

Reaching for the Moon

PROLOGUE

U.S./USSR Early Postwar Rocketry

Soviet rocket builders Sergei Korolev and Valentin Glushko had been rivals for years when they started working together at the end of World War II. Each had been excited by the prospects for rocketry early on. When he was a boy, Glushko had even corresponded with Russian spaceflight godfather Konstantin Tsiolkovsky. Both Glushko and Korolev went on to play key roles in rocket experiments in the 1930s. Korolev became a leading light in the Moscow rocketry organization GIRD (Gruppa Isutcheniya Reaktivnovo Dvisheniya, or Group for Investigation of Reactive Motion) and its successor, RNII (Reaction Propulsion Scientific Research Institute). Glushko worked there as well, but Korolev's ideas gained primacy, and he went on to build the RP-318, the Soviet Union's first rocket-propelled aircraft. Glushko worked in Korolev's shadow, his bullheadedness and egotism feeding his resentment at Korolev's success.

However, in 1938, at the peak of Joseph Stalin's purges, and before the RP-318 aircraft could make a rocket-propelled flight, Glushko and Korolev, along with other aerospace engineers, were imprisoned in the Gulag. Both had been denounced by two leaders of the RNII, Ivan Kleymenov and Georgy Langemak, for deliberately slowing the institute's work, and for anti-Bolshevik sympathies. Glushko played a role in Korolev's incarceration: in his testimony Glushko, probably under duress, turned on Korolev. Korolev did the same in his testimony. Both went to the Butyrka prison, and Korolev was forced to work for Glushko. Their mutual loathing escalated in that environment, as Stalin's "Great Terror" took its toll. Both Glushko and Korolev suffered in the Gulag, but when World War II began, they were sent to work camps to develop new military weapons. In 1941 Glushko was placed in charge of a design bureau for liquid-fueled rocket engines. Working together in 1944, even as their mutual antipathy continued, Korolev and Glushko designed the RD-1 KhZ auxiliary rocket motor used on a Lavochkin La-7R interceptor.

Late in World War II, both Glushko and Korolev perceived the immense progress that Wernher von Braun's team had made with the V-2 ballistic missile in Nazi Germany. Beginning in 1944, Germany used the V-2 to rain down some 3,172 warheads on Western Europe. The success of this missile led Glushko and Korolev to approach Stalin and champion the development of their own ballistic missiles. In 1945 Stalin sent both men—Glushko was the more trusted of the two—along with other technical experts from the work camp, to Soviet-occupied Germany to investigate Nazi ballistic missile efforts. At first Korolev was relegated to overseeing hardware salvage while Glushko led efforts to interview dozens of V-2 engineers and technicians who still remained in Germany, but soon Korolev began assisting with this as well. He quickly earned the trust of the minister of armaments, Dmitri Fedorovich Ustinov.

When Stalin signed the decree initiating development of Soviet ballistic missiles, on May 13, 1946, Ustinov appointed Korolev director of the newly established Scientific Research Institute NII-88

Figure 1. Sergei Korolev led the Soviet Union's efforts into space until his death in January 1966.

to conduct ballistic missile development. He also appointed Glushko chief designer of his own bureau, the OKB 456, which was charged with rocket engine development, a position he held until 1974. Glushko's OKB 456 (later NPO Energomash) would later design the thirty-five–metric ton (340 kN)–thrust RD-101 engine that powered Korolev's early rockets. Glushko also built the powerful RD-170/RD-180 family of liquid-propellant engines that have been used on many rockets around the world to the present day.

The work of both Korolev and Glushko was central to the post–World War II ballistic missile race between the United States and the Soviet Union, as well as the space race to the Moon. Both played critical roles not only in making the dreams of spaceflight real but also in the nuclear terror made possible by intercontinental ballistic missiles (ICBMs) between the 1960s and the 1980s.

Korolev had two hundred German employees of the Mittelwerk V-2 factory rounded up on the night of October 22–23, 1946, and sent to relatively comfortable living quarters at Lake Seliger, between Moscow and Leningrad. The Germans had little direct contact with Korolev's engineers; they assisted in launching a few V-2s from Kapustin Yar. Mainly they answered written questions and were finally returned to East Germany between 1950 and 1954. Korolev copied the V-2 design with Soviet components for his R-1 missile; these evolved during the 1950s into the successively more capable R-2 and R-5 missiles. Korolev received approval for development of the R-7 on May 20, 1954 (although preliminary ICBM development was approved while Stalin was alive, on February 13, 1953). Again, Glushko built the engines that powered this missile.

The Soviets successfully tested the R-7 on August 21, 1957, a two-stage ICBM with a maximum payload of 5.4 tons, sufficient to carry a Soviet atomic warhead thirty-five hundred miles. While it was the world's first ICBM, the R-7 was an impractical weapon. It required enormous launch pads, complex assembly and launching procedures, cryogenic liquid-oxygen oxidizer, and radio-controlled terminal guidance. Moreover, its range was impressive but sufficient to reach only the far northern United States from a launch site in the Soviet Union. As a result, the Red Army deployed it as a weapon at only eight launch pads in Tyuratam and Plesetsk, in the northern USSR. Development of more practical successors, such as Korolev's R-9, did not begin until May 13, 1959. The R-7 served well as an early space-exploration vehicle, however, and it launched both *Sputnik 1* and 2 into orbit in 1957 and served as the first human space launcher for Yuri Gagarin and Gherman Titov in 1961. As much as anyone, Korolev led the Soviet effort to reach the Moon, with the sometime support of Glushko and the strong rivalry he provided. The death of Korolev in a botched medical procedure in early 1966 may have done more to retard the Soviet Moon program than any other single setback; it certainly left Glushko in a position to assume leadership in the Soviet space program.

An American counterpart to the "frenemy" relationship of Korolev/Glushko may be found in the careers of Robert R. Gilruth and Wernher von Braun. During the latter part of World War II leaders of the National Advisory Committee for Aeronautics (NACA), the predecessor to NASA, had become interested in the possibilities of high-speed guided missiles and the future of spaceflight. NACA officials created at the end of World War II the Pilotless Aircraft Research Division (PARD) under the leadership of Gilruth, then a young and promising engineer at the Langley Memorial Aeronautical Laboratory in Tidewater Virginia.

Gilruth established Wallops Island near the Virginia shore as a test facility under the control of Langley on July 4, 1945. From this site the NACA launched between 1947 and 1949 more than three hundred rockets of all sizes and types, leading to the publication of its first technical report on rocketry, "Aerodynamic Problems of Guided Missiles," in 1947. From this, Gilruth and PARD filled in the gaps in the knowledge of spaceflight. In 1952, for example, PARD started the development of multistage, hypersonic, solid-fuel rocket vehicles. These vehicles were initially used primarily in aerodynamic heating tests and were then directed toward a reentry physics research program. On October 14, 1954, PARD launched the first American four-stage rocket, and in August 1956 it launched a five-stage, solid-fuel rocket test vehicle, the world's first, that reached a speed of Mach 15, fifteen times the speed of sound. These strides in the development of rocket technology positioned the NACA as a quintessential agency in the growing importance in the 1950s of the quest for space.

When the National Aeronautics and Space Administration (NASA) began operations in 1958, Gilruth accepted responsibility for its first signature program, Project Mercury, to place the first astronauts into orbit. Dottie Lee, the only woman engineer working in Gilruth's Space Task Group, remembered him as "this beautiful, brilliant man." "I worked in an office with perhaps seven men, and I'm in my little corner," Lee recalled in an oral history in 1999. Gilruth "stops at the door, and I can see him. The men are discussing something, trying to solve a problem, and he listens. Then he asks a question, which turned their thinking around and headed them down the right path. And he turned around, with a smile on his face, and walked out. . . . He didn't tell them how to do it; he just asked a question. . . . And I thought, 'Why can't everyone be like this man?' "

Gilruth went on to lead the Space Task Group for NASA that accomplished Project Mercury, then served as director of the Manned Spacecraft Center—renamed the Johnson Space Center in 1973—which had suzerainty over Gemini and Apollo. His organization recruited, trained, and oversaw the astronauts and the human spaceflight program throughout the heroic age of spaceflight. Yet his name is much less well known than many others associated with these projects. He was a contemporary on a par with von Braun, the technical director of the Nazi V-2 ballistic missile program, the rocketeer who built the first launcher sending spacecraft into orbit for the United States.

Gilruth later built the NASA center in Houston as the home of the space exploration and became its first director. In this position he pushed hard to develop the human spaceflight program as a cohesive whole during the 1960s. His longtime associate, Chris Kraft, said of him at the time of his death in 2000: "He was the guy with the right thoughts in his mind, with the right kind of leadership, and had sense enough to get the right kind of people to do the right kind of job, and you can't say anything better about a man than that."

Gilruth was an example of the engineering entrepreneur, a developer and manager of complex technological and organizational systems, accomplishing remarkably difficult tasks through excellent oversight of the technical, fiscal, cultural, and social reins of the effort. Johnson Space Center director George W. S. Abbey offered this eulogy about Gilruth's career on NASA's behalf in 2000: "Robert Gilruth was a true pioneer in every sense of the word and the father of human space flight. His vision, energy and dedication helped define the American space program. His leadership turned the fledgling Manned Spacecraft Center into what it is today, the leader in humanity's exploration of outer space."

A counterpoint to the career of Robert Gilruth in the United States was the experience of the handsome German émigré Wernher von Braun, one of the most important rocket developers and champions of space exploration during the period between the 1930s and the 1970s. Raised on the science fiction of Jules Verne and

Figure 2. Robert R. Gilruth, left, and Wernher von Braun, right, led the two most important NASA centers during the space race. Gilruth directed the Manned Spacecraft Center (renamed the Lyndon B. Johnson Space Center in 1973), in Houston, while von Braun directed the George C. Marshall Space Flight Center, in Huntsville, Alabama. Both played critical roles in enabling the American Moon landings.

H. G. Wells, as well as the scientific writings of Hermann Oberth and others, von Braun joined the German rocket society Verein für Raumschiffahrt (VfR) as a teenager in 1929. As a means of furthering his desire to build large and capable rockets, in 1932 he went to work for the German army to develop ballistic missiles, and he worked throughout World War II building the liquid-propelled V-2 missile. His program's use of concentration camp labor from the Dora and Mittelwerk camps raised questions after the war of whether he might have engaged in war crimes.

At forty-six feet in length and weighing twenty-seven thousand pounds, the V-2 flew at speeds greater than thirty-five hundred miles per hour and delivered a twenty-two hundred–pound warhead to a target five hundred miles away. First flown in October 1942, it was used against targets in Europe beginning in September 1944. By the end of the war more than thirty-one hundred V-2s had been launched against Antwerp, London, and other continental targets. The guidance system for these missiles was imperfect, and many did not reach their targets; but they struck without warning, and there was no defense against them. As a result, the V-2 had a terror factor far beyond its capabilities.

By the beginning of 1945 it was obvious to von Braun that Germany would not achieve victory against the Allies, and he began planning for the postwar era. Before the Allied capture of the V-2 rocket complex, von Braun arranged the surrender of his best rocket engineers, along with plans and test vehicles, to the Americans. Because of the intriguing nature of V-2 technology, von Braun and his chief assistants were brought to the United States as part of Project Paperclip. Installed at Fort Bliss, Texas, they worked on rockets for the U.S. Army, launching them at White Sands Proving Ground, New Mexico. In 1950 von Braun's team moved to the Redstone Arsenal near Huntsville, Alabama, where it built the army's Jupiter ballistic missile, a launcher capable of sending a small warhead a maximum of five hundred miles.

Von Braun also became one of the most prominent spokesmen of space exploration in the United States in the 1950s. In 1952 he gained note as a participant in an important symposium dedicated to the subject, and he burst on the nation's stage in the fall of 1952 with a series of articles in *Collier's*, a popular weekly periodical of the era. He also became a household name following his appearance on three Disney television shows dedicated to space exploration in the mid-1950s. He gained a greater status in the 1960s, however, when his rocket team at Huntsville, Alabama, transferred to NASA's Marshall Space Flight Center to build the mighty Saturn V rocket that took astronauts to the Moon.

Gilruth and von Braun, like Korolev and Glushko, worked together during the space race. Each needed the other to be successful, and their relationship was professional; they were sometimes in sync, but also often at odds. Dealings proved complex, rivalries ran deep, and achievements proved astounding. Neither of these leaders could have achieved the Moon landings without the other; a symbiotic relationship evolved as NASA undertook Project Apollo in 1961.

Couple this with the ballistic missile program taking place in the U.S. military services and the stage was set for the beginnings of the space age. Throughout the 1950s all the armed services worked toward the fielding of intercontinental ballistic missiles to deliver warheads to targets half a world away. By the late 1950s, therefore, rocket technology had developed sufficiently for the creation of a viable ballistic missile capability. This was a revolutionary development that gave humanity for the first time in its history the ability to attack one continent from another. It effectively shrank the size of the globe, and the United States, which had been protected from outside attack by two massive oceans, could no longer rely on that natural boundary. Its own capability, additionally, signaled for the rest of the world that the United States could project military might anywhere in the world.

The ICBM program did something more: it helped in the maturation of technologies necessary for the space race. Robert Gilruth probably voiced the ultimate aim of all of these individuals during that period of ferment leading to the space race of the late 1950s through the 1960s. He commented in an oral history: "When you think about putting a man up there, that's a different thing. There are a lot of things you can do with men up in orbit." That goal prompted everything Korolev/Glushko/von Braun/Gilruth did from the 1950s on. It prompted a titanic race to the Moon in which either side might have been first, depending on how certain events turned out. These engineers undertook the building of rockets and other technology necessary to make space exploration a reality. Collectively, these individuals and thousands like them helped to make the dreams of spaceflight real.

ONE

Sputnik Winter

Few Americans considered the reception on Friday, October 4, 1957, at the Soviet Union's embassy in Washington, D.C., to be anything out of the ordinary. It was the appropriate culmination of a weeklong set of international scientific meetings. It was also, in the cynical Cold War world of international intrigue between the United States and the Soviet Union, an opportunity to gather national security intelligence and engage in petty games of one-upmanship between the rivals. This reception, however, would prove far different. The one-upmanship continued, but it was far from petty. To a remarkable degree, the Soviet announcement that evening changed the course of the Cold War.

Dr. John P. Hagen arrived early at the party; he wanted to talk to a few Soviet scientists, those he considered personal friends from long years of association in international scientific organizations, to learn their true feelings about efforts to launch an artificial satellite as part of the International Geophysical Year (IGY). Hagen, a senior scientist with the Naval Research Laboratory, headed the American effort on Project Vanguard, and the rocket was behind schedule and over budget. Was the same true of the Soviet Union, or would the satellite go up in 1958 as planned?

Hagen had been through a wringer in the past week. Beginning on Monday, September 30, the international scientific organization known as CSAGI (Comité Spécial de l'Année Géophysique Internationale) had opened a six-day conference at the National Academy of Sciences in Washington on rocket and satellite research for the IGY. Scientists from the United States, the Soviet Union, and five other nations were meeting to discuss their individual plans and to develop protocols for sharing scientific data and findings. Hints from the Soviets at the meeting, however, had thrown the conference into a tizzy of speculation. Several Soviet officials had intimated that they could probably launch their scientific satellite within weeks instead of months. Hagen worried that the offhand remark on the conference's first day by scientist Sergei M. Poloskov that the Soviet Union was "on the *eve* of the first artificial earth satellite" was more than boastful rhetoric. What would a surprise Soviet launch mean for his Vanguard program and for the United States, Hagen wondered.

Hagen did not have long to wait to learn the answer. The party had gathered in the second-floor ballroom at the embassy when a little before 6:00 P.M. Walter Sullivan, a science reporter with the *New York Times* who was also attending the reception, received a frantic telephone call from his Washington bureau chief. Sullivan learned that the Soviet news agency Tass had just announced the launch of *Sputnik 1*, the world's first Earth-orbiting artificial satellite. When he returned to the party, Sullivan sought out Richard W. Porter, a member of the American IGY committee, and whispered, "It's up." Porter's ruddy face flushed even more as he heard this news, although he too had suspected *Sputnik*'s imminent launch. He glided through the gaggles of scientists, politicians, journalists, straphangers, and spies in search of Lloyd V. Berkner, the official American delegate to CSAGI.

When told the news Berkner acted with the characteristic charm of his polished demeanor. Clapping his hands for attention, "I wish to make an announcement," he declared. "I've just been informed by the *New York Times* that a Russian satellite is in orbit at an elevation of 900 kilometers. I wish to congratulate our Soviet colleagues on their achievement." On the other side of the ballroom Hagen's face turned pale. They had beaten the Vanguard satellite effort into space. Were they really the greatest nation on Earth, as Soviet leaders boisterously reminded anyone who would listen? Were they really going to bury us, as Soviet Premier Nikita Khrushchev had announced at the United Nations in 1960, as he pounded his fist and then his shoe on his desk? What could the United States do to recover a measure of international respect?

At the IGY reception the scientists immediately adjourned to the Soviet embassy's rooftop to view the heavens. They were not able to see the satellite with the naked eye. Indeed, *Sputnik 1* twice passed within easy detection range of the United States before anyone even knew of its existence. The next morning at the IGY conference, the Soviet Union's chief delegate, Anatoli A. Blagonravov, explained details of the launch and the spacecraft. The CSAGI conference officially congratulated the Soviets for their scientific accomplishment. But what was not said, but clearly thought by many Americans in both the scientific and political communities, was that the Soviet Union had staged a tremendous propaganda coup for the communist system, and that it could now legitimately claim leadership in a major technological field. The international image of the Soviet Union was greatly enhanced overnight.

The inner turmoil that Hagen felt on "Sputnik Night," as October 4–5, 1957, has come to be called, reverberated through the American public in the days that followed. Two generations after

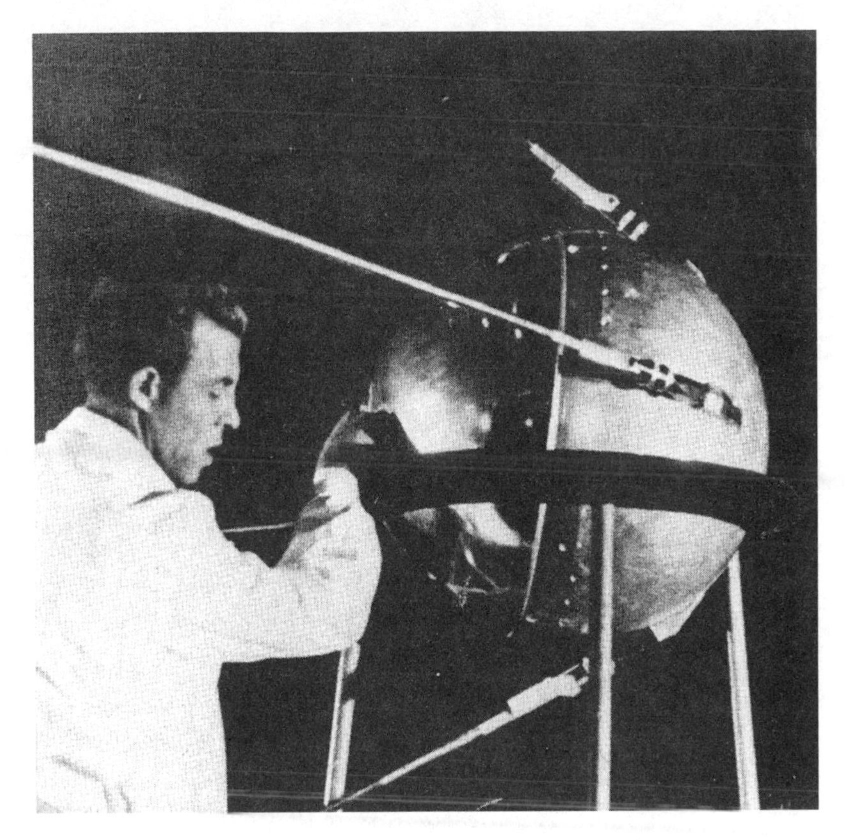

Figure 3. A Soviet engineer completes some last-minute work on the *Sputnik 1* spacecraft, launched by the Soviet Union on October 4, 1957.

the event, words do not easily convey the American reaction to the Soviet satellite. The only appropriate characterization that begins to capture the mood on October 5 demands the use of the word *hysteria*. A collective mental turmoil and soul-searching followed, as American society thrashed around for the answers to Hagen's questions. Almost immediately, two phrases entered the American lexicon to define time, "pre-Sputnik" and "post-Sputnik." The other phrase that soon replaced earlier definitions of time was "Space Age," for with the launch of *Sputnik 1* the space age had been born and the world would be different ever after.

The Nature of Cold War Competition

The *Sputnik* launch created such broad repercussions because of the long-standing difficult relations between the United States and the Soviet Union. From the time of the Bolshevik Revolution in October 1917 that brought Lenin and the communists to power in Russia, the United States had resisted the creation of a state that sought to level society through wealth redistribution. American troops intervened to support more moderate Russian forces in the revolution. These efforts failed, and American troops finally evacuated Russia only in 1920. The United States also refused to acknowledge the legitimacy of the Soviet government until November 16, 1933, when President Franklin D. Roosevelt ended almost sixteen years of American nonrecognition of the Soviet Union following a series of negotiations in Washington, D.C., with the Soviet commissar for foreign affairs, Maxim Litvinov.

This active resistance by the United States to the establishment of the Soviet state created an environment of distrust. The Cold War that coalesced in the late 1940s rested on this long-standing situation and proved much more than just a military standoff. This rivalry between the United States and the Soviet Union manifested itself in an all-out competition between two divergent political and economic systems. While the United States and the Soviet Union were allied in World War II to defeat the Axis Powers led by Nazi Germany, that alliance rested on the core understanding that "the enemy of my enemy is my friend." No sooner had the Allies won that war than the nations returned to an uneasy relationship.

After the war, these two rival nations nearly came to blows, and by that time each had nuclear weapons that could annihilate the other. The rivals squared off over the future of Central and Eastern

Europe beginning in the late 1940s and lasting until the early 1990s. Accordingly, for nearly four decades the Soviets and the Americans squared off in this Cold War. Virtually every place in the world was a potential flashpoint; the United States sought to "contain" the spread of communism outside of the areas controlled by the USSR at the end of World War II. In July 1947, the quarterly journal *Foreign Affairs* published an anonymous article entitled "The Sources of Soviet Conduct," which advocated a strategy of containment in dealing with the Soviet Union. Its author, soon revealed as U.S. State Department official George F. Kennan, proposed active opposition to any expansion of communist power. Kennan wrote, "We are going to continue for a long time to find the Russians difficult to deal with. It does not mean that they should be considered as embarked upon a do-or-die program to overthrow our society." Place Stalin, leader of the Soviet Union, in a box, and limit his ability to do anything internationally.

Containment became America's official strategy, and in seeking to accomplish it the United States forged numerous alliances in Western Europe, the Middle East, and Southeast Asia. The strategy, however, engendered many controversies, some of which verged on erupting into full-scale wars. The Berlin blockade (1948–1949), the Korean War (1950–1953), the Vietnam War (1959–1975), the Cuban missile crisis (1962), and the Soviet-Afghan War (1979–1989) all served to heighten tensions. In at least one instance, the Cuban missile crisis, nuclear holocaust almost came to pass. There were also periods when tensions declined, especially during the détente of the early 1970s, when the nations opened more amiable relations. In the end, direct military actions were deterred by the potential for mutual assured destruction, MAD—certainly the most appropriate acronym ever dreamed up, since it referred to annihilation of both combatants through the use of nuclear weapons.

By the time of the launch of the world's first artificial satellite in

1957, the United States and the Soviet Union each faced critical strategic challenges. Both were consumed by fears of an advance of the other into territories. The United States and its allies were on the watch for Soviet advances in Europe and Asia. The Soviet Union and its allies considered themselves surrounded by hostile nations led by an aggressive United States. The Soviets, moreover, had no bases close enough to American soil to strike should war break out, yet they felt threatened by a plethora of U.S. military capabilities based in Europe and Asia. Not until the rise of ballistic missiles would this change.

Depending on how one viewed this geopolitical confrontation, either side looked surrounded. A polar projection map suggested that the Soviet Union was hemmed in by unfriendly forces, most of which were allies of the United States. Europe to the west, various Asian nations to the east, and, after 1952, NATO allies such as Turkey to the south all created nervousness in the Kremlin. That feeling of being boxed in proved psychologically powerful to Soviet leaders and helped to prompt a series of aggressive responses. In the early ballistic missile era of the late 1950s, moreover, the United States placed nuclear weapons on Jupiter missiles in Turkey and Italy and based strategic bombers in other parts of Western Europe and in Japan and South Korea. The seeming paranoia of the Soviet leaders, especially Joseph Stalin and his successor Nikita Khrushchev, may be partially explained by this sense of being crowded by hostile powers. They forever seemed to be seeking breathing room and buffer territory that could make the Soviet Union invulnerable to attack.

A world map in a more common Mercator projection shows an expansion of the Soviet Union as a result of World War II. It depicts the belligerence of Stalin in subsuming Eastern Europe under Soviet control. Out of the war the Soviets incorporated into the Russian sphere of influence much of Eastern Europe, including Poland, Czechoslovakia, Romania, East Germany, and other smaller nations. Struggles in Greece and other parts of Europe narrowly turned back communist takeovers. A theory of foreign relations emerged from this setting that was called the domino effect; its proponents asserted that if one part of a region came under the influence of communism, then surrounding areas might well follow from revolutionary efforts sparked by the neighboring communist state. President Dwight D. Eisenhower accepted this theory, stating on April 7, 1954, "You have a row of dominoes set up, you knock over the first one, and what will happen to the last one is the certainty that it will go over very quickly. So you could have a beginning of a disintegration that would have the most profound influences." Successive U.S. presidents applied this idea to Soviet relations, justifying American intervention around the world, in dealing with presumed threats from the Soviets.

At the same time, playing on the fears Americans had felt for years of a Soviet threat to their way of life, the development of highly sophisticated nuclear weapons raised the stakes in this confrontation. And the Soviets felt the same way about the Americans. The United States had developed the first atomic bomb in World War II, and had used it on Hiroshima and Nagasaki in August 1945 not only to force Japan's unconditional surrender but also to demonstrate this devastating capability for Stalin. The Soviets followed by exploding their own atomic warhead in 1949. The Americans upped the ante in 1952 with the testing of the much more powerful hydrogen thermonuclear device, but the Soviets followed with their own test of a hydrogen weapon in 1953. It appeared as if the Soviet Union was overtaking the United States in the development of new and ever more powerful weapons.

The launch of *Sputnik* in 1957 suggested that Americans had lost out to Soviet technology. Investment banker and Republican operative Frank Altschul wrote to President Eisenhower on this issue on October 8, 1957, outlining the loss of atomic supremacy, general deterioration in the position of the non-Soviet world, successful activities by the Soviet Union in the Middle East, crushing of anti-communist uprisings in Hungary and Czechoslovakia, communist forays into Southeast Asia, and a host of other apparent setbacks for the West. All of this created world doubt about the outcome of the Cold War for democratic nations in general and the United States as its leader in particular. *Sputnik* was the final straw in a long series of problems. Altschul suggested that "the impression of impotence created by the failure of the Western world to find practical measures to counteract overt acts of Soviet aggression" made this a crisis.

The International Geophysical Year and the Origins of the Space Age

The International Geophysical Year (IGY) served magnificently as a vehicle for the birth of the space age. The genesis of the IGY was a dinner party in the home of scientist James A. Van Allen in Bethesda, Maryland, in the summer of 1950. This event has taken on legendary status as everything from a nearly mystical gathering to the reaffirming of the authority of science in modern life. At some level the IGY may be viewed as a cabal led by Van Allen, British physicist Sydney Chapman, and American science entrepreneur Lloyd V. Berkner to "hoodwink" the nations of the world into pursuing an aggressive scientific program to ensure funding and status for activities never dreamed of before. At another, it may be interpreted as cagey politicians manipulating the scientific community to provide a stalking horse for the resolution of a thorny geopolitical problem. Indeed, it is both of these.

Pressed by Berkner, the International Council of Scientific Unions (ICSU) agreed in 1952 to pursue a comprehensive series of global geophysical activities to span the period July 1957—December

1958. Sixty-seven nations agreed to conduct cooperative experiments to study solar-terrestrial relations during a period of maximum solar activity in 1957–1958. In October 1954, at the behest of essentially this same group of U.S. scientists, ICSU challenged nations to use missiles being developed for war to launch scientific satellites to support the IGY research program.

In response to ICSU's announcement, on May 26, 1955, the National Security Council (NSC), the senior defense policy board in the United States, approved a plan to orbit a scientific satellite as part of the IGY effort. The NSC's endorsement was provisional: the effort could not interfere with the development of ballistic missiles, it must emphasize the peaceful purposes of the endeavor, and it had to contribute to establishing the principle of "freedom of space" in international law. Eisenhower supported this effort and on July 29 publicly announced plans for existing organizations within the Department of Defense to develop and launch a small scientific satellite, "under international auspices, such as the International Geophysical Year, in order to emphasize its peaceful purposes[;] . . . considerable prestige and psychological benefits will accrue to the nation which first is successful in launching a satellite . . . especially if the USSR were to be the first to establish a satellite."

There followed a heady competition between the Naval Research Laboratory on the one hand and the army's Redstone Arsenal, where von Braun led the effort to win permission to develop the IGY satellite. Project Vanguard, proposed by the navy, was chosen on September 9, 1955, to carry the standard in launching a nonmilitary satellite for the IGY effort, over the army's "Explorer" proposal. The decision was made largely because the Naval Research Laboratory candidate did not interfere with high-priority ballistic missile programs—it used the Viking launcher as its basis rather than a ballistic missile—while the army's bid was heavily involved in those activities and proposed adapting a ballistic missile for the purpose. In addition, the navy rocket seemed to have greater promise for scientific research because of a larger payload capacity.

The Viking launch vehicle was also a proven system; an early version of it had first flown in the late 1940s, while the army's proposed rocket, the Redstone, had been launched for the first time only in August 1953. Finally, the Naval Research Laboratory's proposal was more acceptable because it came from a scientific organization rather than from a weapons developer, in this case the Redstone Arsenal.

Although he approved the IGY satellite, Eisenhower was cost conscious about the program, especially as it seemed to grow in cost and complexity with every review. He repeatedly wondered about its voracious appetite for public funds, especially since Vanguard supposedly took a back seat to real national security space activities, most notably the accelerating program to develop ballistic missiles. From its initial cost estimates, Vanguard had mushroomed to a cost of $67.9 million by August 1956 and to $110 million by the summer of 1957.

During the next several months the Eisenhower administration became increasingly concerned with the tendency of Project Vanguard to get bogged down. Eisenhower was especially concerned about the probability that the scientific instruments were slowing it down. About five months before the Soviets orbited *Sputnik 1*, the president reminded his top advisers, as his aide Andrew Goodpaster recorded, "Such costly instrumentation had not been envisaged," and the president "stressed that the element of national prestige . . . depended on getting a satellite into its orbit, and not on the instrumentation of the scientific satellite." Eisenhower's perception of the budgetary growth of the Vanguard program, transforming it from the simple task of putting any type of satellite into orbit into a project of launching a satellite with "considerable instrumentation," reminded him of the worst type of technological inflation, as every scientist seemingly wanted to hang another piece of equipment on the vehicle.

Von Braun and his Huntsville rocket team truly believed that they should have received the mandate to build and launch the first American satellite. Having lost to the Vanguard program, however, they pushed for their appointment as a backup plan in case Vanguard was unsuccessful. As the Vanguard program ran into technical difficulties, furthermore, von Braun pressed for permission to launch ahead of the rival effort, claiming he could place a satellite in orbit as early as January 1957. The Department of Defense turned down that request, making the announcement in a memorandum of July 5, 1956:

> While it is true that the VANGUARD group does not expect to make its first satellite attempt before August 1957, whereas a satellite attempt could be made by the Army Ballistic Missile Agency as early as January 1957, little would be gained by making such an early satellite attempt as an isolated action with no follow-up program. In the case of VANGUARD, the first flight will be followed up by five additional satellite attempts in the ensuing year. It would be impossible for the ABMA group to make any satellite attempt that has a reasonable chance of success without diversion of the efforts of their top-flight scientific personnel from the main course of the JUPITER program, and to some extent, diversion of missiles from the early phase of the re-entry test program. There would also be a problem of additional funding not now provided.

The final reason for disapproving this action: "The obvious interference with the progress of the JUPITER program would certainly present a strong argument against such diversion of scientific effort."

Similar problems plagued the Soviet scientific satellite effort. Inspired by the Cold War of the 1950s, the eventual Soviet success had been laboriously built by Sergei Korolev and Valentin Glushko through years of systematic, command-economy efforts to achieve a perceived technological edge in military capability. The rocket received a formal go-ahead in the middle of 1954. The missile was meant primarily to carry nuclear warheads, but Korolev designed it so that it could instead orbit a satellite of 1.5 tons. The launching of a satellite, though unmentioned in planning discussions, was never far from the thinking of those involved in developing the technology.

The Soviet Union officially entered the satellite sweepstakes on August 2, 1955, when it responded to the Eisenhower administration's announcement to launch a satellite during the IGY. At a press conference at the Soviet embassy in Copenhagen before about fifty journalists, Leonid Ivanovich Sedov stated: "In my opinion, it will be possible to launch an artificial Earth satellite within the next two years." He added, "The realization of the Soviet project can be expected in the near future. I won't take it upon myself to name the date more precisely."

On August 30, 1955, Korolev discussed an IGY satellite in the highest levels of the Soviet defense establishment. He received approval to pursue the effort in no small measure because it would be a powerful demonstration of Soviet science and technology during the IGY. From there he went to another meeting in Moscow with the scientific community, including Glushko, where he told them: "As for the booster rocket, we hope to begin the first launches in April–July 1957 . . . before the start of the International Geophysical Year." The IGY schedule gave Korolev a timetable not otherwise possible for any orbital satellite efforts. Within days they had a program fleshed out to undertake study of the ionosphere, cosmic rays, Earth's magnetic fields, luminescence in the upper atmosphere, the Sun and its relationship to Earth, and other natural phenomena. The Soviet Academy of Sciences embraced this program; it would well satisfy the scientific objectives of the IGY and project Soviet capability on the world stage.

The Soviet effort gained official reality on January 30, 1956, when the USSR Council of Ministers issued decree number 149-88ss. The document approved a launch in 1957 of an artificial satellite, designated "Object D," as part of the IGY. Responsibility for the project rested with Korolev's Design Bureau, OKB-1, with support from Glushko and others. Korolev met with Soviet Premier Nikita Khrushchev in February 1956 at OKB-1 to solidify support. While showing Khrushchev the massive R-7 rocket, Korolev turned the conversation to all the many objectives that could be met with the rocket, among them the IGY satellite launching. When it came time for Glushko to speak, he bored Khrushchev with extraneous technical details, "like he was talking to first course students at the neighboring forestry institute . . . rather than the higher leadership." Accordingly, Korolev invoked the name of sainted Russian space theorist Konstantin Tsiolkovsky and told Khrushchev that the Soviet people were on the verge of realizing his dreams of space exploration with the R-7 rocket.

Korolev showed a model of Object D and suggested to the Soviet premier that the Americans were focused on the IGY satellite effort and that this represented an opportunity to best their efforts. The R-7 was nearly ready for launch, he insisted, and its capabilities were such that it could hoist Object D into orbit with energy to spare. In the end, Khrushchev hesitantly told Korolev: "If the main task"— ballistic missile development—"doesn't suffer, do it." With this meeting, the Soviet IGY satellite project was under way.

This decision, however, did not necessarily mean that Korolev would receive all the funding he requested. He constantly grumbled about the need for additional resources. His close friend Mstislav Vsevolodovich Keldysh made the case in a September 14, 1956, meeting with the Presidium of the Soviet Academy of Sciences. Keldysh excited the members with the possibility of placing a dog in orbit, as well as hinting of future human missions. "We, of course, can't stop at the task of creating an Earth satellite. We, naturally, are thinking of further tasks—of space flight. The first project along these lines, I believe, will be to fly around the Moon and photograph it from the side which is always hidden to us." To do this, he insisted, "it would be good if the Presidium were to turn the serious attention of all its institutions to the necessity of doing this work on time. . . . We all want our satellite to fly earlier than the Americans'."

On August 21, 1957, the Soviet Union launched a successful R-7 rocket, its sixth attempt, carrying a dummy warhead more than four thousand miles to Kamchatka. Korolev was convinced by this success that the launcher was ready for use. Object D, however, lagged in development. He then substituted a less complex satellite, a spherical object weighing only 183 pounds, with a radio transmitter, batteries, and unsophisticated measuring instruments. This became known to the world as *Sputnik 1*, "fellow traveler" in Russian.

In a 1993 interview, Oleg Ivanovsky, who had worked on the program, recalled some of the problems with the satellite:

> We had to find new techniques of manufacturing the surfaces in order to achieve the necessary optical and thermal qualities. We had no experience in this work. . . . Korolev, with his iron character, was able to influence the attitude of people. The Party directed that new paint be put on the factory walls. Korolev put the satellite on a special stand, draped in velvet, in order that the workers would show reverence towards it. He supervised the carrying out of the production schedule every day personally.

Korolev became a man obsessed. He "insisted that both halves of the sputnik's metallic sphere be polished until they shone, that they be spotlessly clean," recalled Konstantin Petrovich Feoktistov, who would be the first engineer-cosmonaut to go into orbit in the three-man *Voskhod* seven years later. "The people who developed the radio equipment were actually the ones demanding this," he added. "They were afraid of the system overheating, and they wanted the orbiting sphere to reflect as many rays of the Sun as possible."

Sputnik 1 launched atop an R-7 rocket on October 4, 1957. Red Army Colonel Mikhail Fyodorovich Rebrov recalled the experience:

> No one will ever know what was going through Sergey Korolev's mind at the time. Later on, when the sputnik was installed in orbit, and its call sign was heard over the globe, he said: "I've been waiting all my life for this day!" The moment of the blast-off has been described many times. Then the rocket got out of the radio zone. The communication with the sputnik ended. The small room where the radio receivers were was overcrowded. Time dragged on slowly. Waiting built up the stress. Everyone stopped talking. There was absolute silence. All that could be heard was the breathing of the people and the quiet static in the loudspeaker. . . . And then from very far-off there appeared, at first very quietly and then louder and louder, those "bleep-bleeps" which confirmed that it was in orbit and in operation. Once again everyone rejoiced. There were kisses, hugs and cries of "Hurrah!" The austere men, who were greeted out of space by the messenger they had made, had tears in their eyes.

American Responses to *Sputnik*

Some have called the Soviet launch of *Sputnik 1* "the shock of the century." But that shock only slowly reverberated through the

American public in the days that followed. Most Americans seemed to recognize that the satellite did not pose a threat to the United States, so a spirit of congratulations prevailed, and many people seemed excited by the Soviet success. At the same time, Eisenhower acknowledged the need to "take all feasible measures to accelerate missile and satellite programs." He also moved to assure the American public that all was well, largely succeeding in doing so during the month of October 1957.

Instead of feeling threatened, a generation of Americans seemed to embrace the dawn of the space age as a symbol of progress and a better future both on Earth and beyond. A generation of Americans had been raised on visions of human colonies on the Moon and Mars, great starships plying galactic oceans, and prospects of a bright, limitless future beyond a confining, overcrowded, and resource-depleted Earth. Entertainment leaders like Walt Disney joined Wernher von Braun and other rocketeers in inspiring the embrace of a promising future in space. Taught in the early 1950s that spaceflight loomed just on the cusp of reality, the public now saw that perception coming true. As an example of how one American responded to *Sputnik*, fourteen-year-old Homer Hickam recalled watching "the bright little ball, moving majestically across the narrow star field between the ridgelines" over his home in Coalwood, West Virginia. Hickam said he was inspired to become an aerospace engineer and devote his life to the quest for space. He was just one of many, not only in the United States and the USSR but across the world.

In fact, the best evidence suggests that excitement about prospects for the future dominated the thinking of the American public immediately after the *Sputnik* launch. Three days later, social anthropologist Margaret Mead and her partner Rhoda Métraux began collecting data gauging American responses to *Sputnik*. They asked colleagues and friends around the country to conduct surveys asking three open-ended questions among divergent age, gender, race, economic, and social groups:

1. What do you think about the satellite?
2. How do you explain Russia's getting their satellite up first?
3. What do you think we can do to make up for it?

Mead and Métraux collected 2,991 adult responses by October 18. Few respondents said that the Soviet launch was an unexpected event; an even smaller number registered no knowledge of the launch. As one investigator summarized in a report on this study, "It seems that most informants in the 'Emergency Survey,' whether or not they possessed prior knowledge about artificial satellites, had taken the news of Sputnik in stride and developed a logical, rather than emotional, approach to the topic by the time they were interviewed."

This assessment squares with more scientific analysis of the *Sputnik* response. As a government study reported in October 1958:

> Interpretations of the sputnik's significance likewise show that public concern was not great. Gallup found that only 50 percent of a sample taken in Washington and Chicago regarded the sputnik as a blow to our prestige. Sixty percent said that we, not the Russians, would make the next great "scientific" (actually technological) advance. A poll by the *Minneapolis Star and Tribune* found that 65 percent of a sample in that State thought we could send up a satellite within 30 days following the Russian success, a statistic which included 56 percent of the college-educated persons asked. In the sample of the Opinion Research Corporation, 13 percent believed that we had fallen behind dangerously, 36 percent that we were behind but would catch up, and 46 percent said that we were still at least abreast of Russia.

There is good reason to believe that the response to *Sputnik* was a political construct. George Reedy, a Democratic strategist, wrote to Senator Lyndon B. Johnson of Texas on October 17, 1957, about how the *Sputnik* issue could be used to the party's advantage: "The issue is one which, if properly handled, would blast the Republicans out of the water, unify the Democratic Party, and elect you President." Reedy suggested, "You should plan to plunge heavily into this one. As long as you stick to the facts and do not get partisan, you will not be out on any limb."

Using every tool at their disposal, Johnson and his associates worked to maximize the *Sputnik* launch for their political purposes. Speaking for many Americans, he remarked in two speeches in Texas in the fall of 1957 that the "Soviets have beaten us at our own game—daring, scientific advances in the atomic age." Since those Cold War rivals had already established a foothold in space, Johnson proposed to "take a long careful look" at why the U.S. space program was trailing that of the Soviet Union. He led a broad review of American defense and space programs in the wake of what he presented as the *Sputnik* crisis. Eventually, the public may have grown to fear the ramifications of the satellite.

President Eisenhower and other leaders of his administration congratulated the Soviet Union and tried to downplay the importance of the *Sputnik 1* accomplishment, but they misjudged the public reaction to the event. The Democrats accused the Eisenhower administration of letting the Soviet Union best the United States. The *Sputnik* crisis reinforced for many people the popular conception that President Eisenhower was a smiling incompetent; it was another instance of a "do-nothing," golf-playing president mismanaging events. With the prodding of Democrats, the public began to perceive *Sputnik* as an illustration of a technological gap, and that perception provided the impetus for increased spending for aerospace endeavors, technical and scientific educational programs, and the chartering of new federal agencies to manage air and space research and development. Not only had the Soviets been first in orbit, but *Sputnik 1* was much larger than the projected 3.5 pounds for the first satellite to be launched by the Americans with Project Vanguard. In the Cold War environment of the late 1950s, this disparity of capability portended menacing implications.

Concerns about Soviet success compounded on November 3, 1957, when the Soviets succeeded in launching a second satellite, and this time it was a monstrosity that carried a dog, Laika. While the first satellite had weighed less than 185 pounds, this spacecraft weighed 1,120 pounds and stayed in orbit for almost two hundred days. Eisenhower tried to quell public apprehensions, and he took action to address the perceived space gap. He accelerated missile programs, put more focus on science and technology in the government, approved a secondary satellite effort called Explorer, and appointed a presidential science adviser. He also created, at the suggestion of scientists in Washington, a President's Science Advisory Committee (PSAC), which began operation on November 22.

As a first tangible effort to counter the apparent Soviet leadership in space technology, the White House announced that the United States would test launch a Project Vanguard booster on December 6, 1957. Media representatives were invited to witness the launch in the hope that they could help restore public confidence, but it was a disaster of the first order. During the ignition sequence, the rocket rose about three feet above the platform, shook briefly, and disintegrated in flames. John Hagen, who had been working feverishly to ready the rocket for flight, was demoralized. He felt even worse after the next test. On February 5, 1958, the Vanguard launch vehicle reached an altitude of four miles and then exploded. Hagen was tearful at the very public failures and some of his associates later thought that his career ended then and there, for he never again held an important post.

In this crisis the army, featuring the handsome and charismatic Wernher von Braun and his rocket team of German immigrants, dusted off an unapproved plan for the IGY satellite effort, Project Explorer, and flew it within an amazingly short period of time. After two launch aborts that made observers apprehensive that the United States might never duplicate the Soviet successes in spaceflight, the *Juno 1* booster carrying *Explorer 1* lifted off from the Cape Canaveral, Florida, launch site at 10:55 P.M. on January 31, 1958. The Juno booster was largely a variant of the ABMA's Jupiter-C, whose technology, in parts, can be traced to the V-2 of World War II. The tracking sites marked the course of the rocket to the upper reaches of the atmosphere, but observers on the ground had to wait to learn whether orbit had been achieved.

Von Braun, who was at the Pentagon with other Department of Defense officials preparing for a press conference, received news from the Cape that the launch had taken place and calculated that telemetry from *Explorer 1* should be received at the West Coast tracking stations at precisely 12:41 A.M. But that time came and went, and still von Braun waited for communication from the satellite. It finally came at 12:49 A.M., when the Jet Propulsion Laboratory (JPL) tracking station confirmed *Explorer 1*'s pass overhead. The delay had amounted to nothing more than a little higher orbit than anticipated and therefore a longer period required to travel the extra mileage.

The spacecraft carried a small instrument, essentially a Geiger counter, to measure radiation encircling Earth. The instrument had been built by James A. Van Allen, a physicist from the University of Iowa. Its data verified the existence of Earth's magnetic field and discovered what came to be called the Van Allen radiation belts. These phenomena partially dictate the electrical charges in the atmosphere and the solar radiation that reaches Earth. Later that day, February 1, 1958, a press conference took place at the National

Figure 4. This iconic image was taken at a press conference at the National Academy of Sciences building in Washington, D.C., after the launch of *Explorer 1* on January 31, 1958. Pictured holding up a model of the rocket are, from left, William Pickering, director of the Jet Propulsion Laboratory and lead on the science effort for *Explorer 1;* James Van Allen, scientific principal investigator for the mission; and Wernher von Braun, technical director of the Army Ballistic Missile Agency, which built and launched the Juno rocket that placed *Explorer 1* into orbit.

Academy of Sciences, where von Braun, Van Allen, and JPL director William H. Pickering announced success. The signature image that appeared in newspapers around the nation the next morning depicts three smiling men holding a full-scale model of *Explorer 1* above their heads in triumph of launching the first United States artificial satellite. Project Vanguard also received additional funding to accelerate activity during this period, and *Vanguard 1* was finally orbited on March 17, 1958, confirming the existence of the Van Allen belts and measuring their severity.

Achieving "Freedom of Space"

The launch of *Sputnik 1* helped to establish for the Eisenhower administration "freedom of space" in international law. Accepted practice—not universally accepted—allowed nations legally to board and confiscate vessels within territorial waters near their coastlines and to force down aircraft flying in their territorial airspace. But space was a territory not defined as yet, and the U.S. position was that it should be recognized as free territory not subject to the normal confines of territorial limits. An opposite position, however, argued for the extension of territorial limits into space above a nation into infinity. Eisenhower pursued a freedom of space accord on July 21, 1955, when he met the Soviets in a Geneva, Switzerland, summit, but they quickly rejected his proposal.

But when *Sputnik* overflew the United States and other nations of the world, it defined a de facto principle of freedom of space, what Eisenhower sometimes referred to as the "Open Skies" doctrine. On October 8, 1957, Deputy Secretary of Defense Donald A. Quarles told the president: "The Russians have . . . done us a good turn, unintentionally, in establishing the concept of freedom of international space." Eisenhower immediately grasped this serendipitous victory as a means of pressing ahead with the launching into orbit of any type of satellite, including those designed for reconnaissance. The pattern held for later satellites, and by the end of 1958 a tenuous "Open Skies" precedent had been established. By happenstance, the Russian space program had set in order the U.S.-backed claim to free access.

Some have speculated that Eisenhower might actually have held back the U.S. effort to launch an orbital satellite to allow the Soviets to do so first, thereby letting them stumble onto this all-important principle of overflight. After all, had the United States launched before the Soviet Union, Khrushchev might have protested the flight as a violation of his nation's airspace. This could have thrown the freedom of space concept into years of intense and confrontational international negotiation. While this is a fascinating possibility, there is no evidence to believe that the Eisenhower administration conspired to lose the race to launch the first satellite. Through serendipity, Eisenhower achieved the right of overflight because of *Sputnik*.

The Birth of NASA

Some of the political pressure on the Eisenhower administration to respond to the Soviet success with *Sputnik* eased with the launch of *Explorer 1*. But not enough to prevent a transformation in the structure of government. As a direct outgrowth of this crisis in the winter of 1957–1958, the administration worked with congressional leaders to draft legislation creating a permanent federal agency dedicated to exploring space. Numerous proposals surfaced during that winter, the least acceptable, at least from Eisenhower's perspective, a plan sponsored by two Democrats, Representative John L. McClellan of Arkansas and Senator Hubert Humphrey of Minnesota, to create a Department of Science and Technology. But Eisenhower resisted other less ambitious plans as well.

A turning point came on February 4, 1958, when he finally capitulated and asked his science adviser, James R. Killian, to convene the President's Science Advisory Committee to come up with a plan for a new spaceflight organization. Quietly considering the creation of a new civil space agency for several months, PSAC worked with staff members from Congress and quickly came forward with a proposal that placed all nonmilitary efforts relative to space exploration under a strengthened and renamed National Advisory Committee for Aeronautics.

Established in 1915 to foster aviation progress in the United States, the NACA had long been a small, loosely organized, and elitist organization known for both its technological competence and its apolitical culture. It had begun moving into space-related areas of research and engineering during the 1950s, through the work of an organization under the leadership of Robert R. Gilruth. Although it was a civilian agency, the NACA also enjoyed a close working relationship with the military services, helping to solve research problems associated with aeronautics and also finding application for them in the civilian sector. Its civilian character, its recognized excellence in technical activities, and its quiet, research-focused image all made it an attractive choice. It could fill the requirements of the constrained job Eisenhower envisioned without exacerbating Cold War tensions with the Soviet Union.

Eisenhower accepted PSAC's recommendations and sponsored legislation to expand the NACA into an agency charged with the broad mission to "plan, direct, and conduct aeronautical and space activities," to involve the nation's scientific community in these activities, and to widely disseminate information about these activities. An administrator appointed by the president was to head the National Aeronautics and Space Administration (NASA). During the summer of 1958 Congress passed the National Aeronautics and Space Act, and the president signed it into law on July 29, 1958. This ended the debate over the type of organization to be created, and other plans died a quiet death. The new organization started functioning on October 1, 1958, less than a year after the launch of *Sputnik 1*. Its first task became the development of a human space-exploration program, Project Mercury. NASA has continued to direct the human space-exploration initiatives of the United States ever since.

To establish a clear direction for the agency, NASA's leadership staff developed a ten-year plan for space exploration. It emphasized the scientific and technological developments to be attained through the effort in each of the following areas: space vehicle development, human spaceflight, engineering and scientific research, and spaceflight operations. The ten-year program called for an expenditure of approximately $12.5 billion in 1959 dollars (equivalent to more than $100 billion in 2019) to accomplish a hefty scientific probe program and a human spaceflight program that would launch its first astronaut in 1961 and achieve a lunar landing at some unspecified time in the post-1970 period. It also provided for the development of new launch boosters that would give the United States a decided edge in long-term space activities. It was a modest but reasonable and optimistic program, and the Eisenhower administration accepted the funding priorities. NASA's first administrator, T. Keith Glennan, wrote in his diary, "Ike and I agreed that we were mature enough as a nation not to let some other country determine our behavior and policy. Hence, we opposed a 'Space Race,' and while we wanted to advance rapidly, not to do foolish things just because the Russians were doing them." At the same time, Glennan recognized that the United States could not operate in a "business as usual" mode while the Soviets were influencing world opinion in their favor by executing "space spectaculars." He told the president in November 1959:

> Personally, I do not believe we can avoid competition with the Soviets in this field. I do not believe we should want to avoid that competition. But I do believe that we can and should establish the terms on which we are competing. We could thus place the "Space race" in proper perspective with all the other activities in the competition between the US and USSR. In doing this, thoughtful consideration should be given to the

> particular and dramatic role occupied by space activities in the whole gamut of international competition.

Glennan opted for a deliberate program with clear objectives and a lengthy timetable.

TWO

The First Race to the Moon

Between 1958 and 1960 the United States went head-to-head with the Soviet Union in a robotic race to the Moon—and lost. The lure of the Moon was irresistible. In 1958, eager to demonstrate leadership in space technology, the United States started an expedited effort to send a series of spacecraft named *Pioneer* to the Moon. The air force prepared three and the army two Pioneer spacecraft for the Moon flights. During the winter of 1958–1959 the United States made four attempts to send a Pioneer probe to the vicinity of the Moon. None reached its destination. Indeed, none succeeded in escaping Earth's orbit, but two produced the first information about the outer regions of the Van Allen belts. In contrast, after some false starts in the fall of 1958, the Soviet Union succeeded in launching several successful probes to the Moon.

This success rested on the early capability of Soviet engineers under the leadership of Korolev and Glushko to build large rockets with significant payload capacity, something not yet accomplished in the United States. In January 1959 the Soviets sent *Luna 1* past the Moon and into orbit around the Sun, following up with *Luna 3* to transmit pictures of the far side of the Moon—thereby giving the Soviets an important "first" in lunar exploration. American efforts lagged the Soviets', but the work of von Braun and Gilruth proved significant in helping the Americans catch up. Eventually, *Pioneer 5* finally flew past the Moon, much too late to assuage America's loss of pride and prestige. Thus ended the first phase of lunar exploration, with the Soviet Union a clear winner.

Dreams of Luna

What is it about the Moon that captures the fancy of humankind? A silvery disk hanging in the night sky, it conjures up images of romance and magic. It has been credited with foreshadowing important events, both good and ill, and its phases for eons served as humanity's most accurate measure of time. Since ancient times, people have watched the Moon wax (appear to grow larger) and wane (appear to shrink), and have wondered at its beauty and mystery. The Moon holds an important place in many of the world's religions, and once had a part in other religions—such as Christianity—that no longer assign it special significance. Many religions have seen the Moon as a deity, with many names and many incarnations.

The Moon is by far the most dominant and changeable element in the night sky. It has kindled enthusiasm, joy, lust, fear, and horror upon generations of peoples of all races and cultures who have lived out their lives under its silvery reflected light. Defining the Moon differently from culture to culture and age to age, humankind remains captivated by its power. We have characterized it by its features, by its phases, and by its influence over earthly entities. Moongazing remains one of the oldest pastimes in the human experience. Ancient civilizations assigned the Moon dominion over human lives through supernatural intervention; more recent ones have envisioned it as a home for extraterrestrial life. It inspires poets and artists, scientists and engineers, creators and destroyers. With the invention of the telescope at the turn of the seventeenth century—coinciding with the rise of the scientific revolution—the Moon took on new meaning as a tangible place with mountains and valleys and craters that could be named, geological features and events that could be studied.

With the scientific revolution, as the Moon came to be considered a place with firm ground, many people began to speculate on the potential of visiting. Johannes Kepler, a pathbreaking astronomer in the seventeenth century, wrote a novel, *Somnium* (Dream), that was published posthumously in 1634. The work recounts a dream of a supernatural voyage to the Moon in which the visitors encounter serpentine creatures. Kepler also included much scientific information in the book, speculating on the difficulties of overcoming Earth's gravitational field, the nature of the elliptical paths of planets, the problems of maintaining life in the vacuum of space, and the geographical features of the Moon.

Other writings sparked by the invention of the telescope and the success of *Somnium* also described fictional trips into space. Cyrano de Bergerac, for example, wrote *Voyage dans la Lune* (The Voyage to the Moon, 1649), describing several attempts by the protagonist to travel to the Moon. First, he ties a string of bottles filled with dew around himself, so that when the heat of the Sun evaporates the dew he will be drawn upward, but he makes it only as far as Canada. Next, he tries to launch a vehicle from the top of a mountain by means of a spring-loaded catapult, "but because I had not taken my measures aright, I fell with a slosh on the Valley below." Returning to his vehicle, Cyrano's hero finds some soldiers mischievously tying firecrackers to it. As they light the fuse, he jumps into the craft and tier upon tier of explosives ignite like rockets and launch him to the Moon. Thus Cyrano's hero becomes the first flyer in fiction to reach the Moon by means of rocket thrust, presaging Newton's third law of gravity, that every action has an equal and opposite reaction. Once on the Moon, Cyrano's protagonist has several adventures, and later he journeys to the Sun.

While these early fictional accounts were not scientifically accurate, later science fiction writers such as Jules Verne and H. G. Wells strove for greater accuracy. Both were well aware of the scientific underpinnings of spaceflight, and their speculations reflected reasonably well what was known at the time about the nature of other worlds. Both Wells and Verne incorporated into their novels a much more sophisticated understanding of the realities of space than had been seen before. Their space vehicles became enclosed capsules powered by electricity and reflecting relatively sound aerodynamic principles. Most of Wells's and Verne's concepts stood up under some, although not all, scientific scrutiny.

For example, in 1865 Verne published *De la Terre à la Lune* (From the Earth to the Moon). The scientific principles informing this book were reasonably accurate for the period, even though Verne took a measure of dramatic license. The novel describes the problems of building a vehicle and launch mechanism to visit the Moon. At the end of the book, Verne's characters are shot into space by a nine hundred–foot–long cannon. Verne picked up the story in a second novel, *Autour de la Lune* (Around the Moon), describing a lunar orbital flight, but he did not allow his characters to land. Likewise, Wells published *The War of the Worlds* in 1897 and *The First Men in the Moon* beginning in 1900, using sound scientific principles to describe space travel and encounters with aliens.

In 1902 French filmmaker Georges Méliès filmed the first story of lunar exploration, *Le Voyage dans la Lune* (A trip to the Moon). Incorporating elements from the fiction of both Wells and Verne, it became a classic almost immediately. An ailing Jules Verne even visited the set. Méliès's film was a highly entertaining story of scientists shot out of a cannon to explore the Moon. They encountered life on the Moon and eventually had to escape back to terra.

Beginning in the 1920s, a space craze gained popularity worldwide as science fiction and science fact reinforced each other to make the dreams of spaceflight seem more real than ever. While meanings shifted significantly over time—in ways that reflected similar shifts in politics, culture, and society—the level of interest and excitement for the possibilities of spaceflight remained constant until the robotic race to the Moon.

The space craze consumed the revolutionary Soviet Union as well. Aleksey Tolstoy's 1923 novel *Aelita, or The Decline of Mars*, was adapted into one of the earliest state feature films. Directed by Yakov Protazanov and made at the Mezhrabpom-Rus studio in 1924, *Aelita* tells the story of Los, who flies to Mars by rocket, meets his love—Queen Aelita—and leads a revolt against the ruling Elders. In the process, they establish on Mars a "workers' state" modeled on the Soviet Union. The early example of a science fiction feature film was popular in the Soviet Union for several years.

At about the same time in Germany, Hermann Oberth published his classic book *Die Rakete zu den Planetenräumen* (The rocket into interplanetary space). The book explained the mathematical theory of rocketry, applied it to possible designs for practical rockets, and considered the potential of space stations and human travel to other planets.

The success of Oberth's 1923 book prompted silent movie maker Fritz Lang to film an adventure story about space travel. The result was the 1929 feature *Die Frau im Mond* (The woman in the Moon). Lang wanted his movie set to be technically correct so he asked Oberth to be his technical adviser. Oberth and science writer Willy Ley helped Lang with his sets and built a spacecraft that looked realistic. Ever the dramatist, Lang even invented the countdown to increase tension for the audience and to add spectacle to the rocket flight. As a publicity stunt for Lang's film, Oberth agreed to build an actual rocket that would be launched at the premier of *Die Frau im Mond*. Two days before the premier, however, Oberth discovered that the rocket could not be completed in time. Regardless, the movie and the envisioned trip to the Moon served as a powerful merging of space technology with imagination.

Other representations of spaceflight also captured the imagination of many people in this era. Capitalizing on the success of Buck Rogers—a fictional space hero in newspaper comic strips, books, films, and memorabilia beginning in the late 1920s—in 1934 Alex Raymond created the *Flash Gordon* comic strip, and a host of toys, games, movies, and other items appeared thereafter.

This mixture of science fiction with science fact illustrates how closely the dominant trends in science fiction literature and film shaped public perceptions and public understanding about space exploration. As reality and perceptions converged, they influenced public expectations that in turn encouraged the pursuit of aggressive space-exploration agendas in both the Soviet Union and the United States, with the Moon as the first target.

Accordingly, as spaceflight became an ever more realistic possibility, the Moon took on added meaning as Earth's nearest astronomical neighbor and a relatively easy destination for humankind to visit and explore. In the desperate rivalry between the United States and the Soviet Union during the Cold War, the Moon held enormous potential as a public relations coup for the nation reaching it first. The number of spaceflight "firsts" associated with the Moon in the 1950s and 1960s clearly demonstrated the significance assigned to lunar exploration during this first heroic era of the space age. From the first clear images of the Moon to the landings of the Apollo astronauts between 1969 and 1972, the Moon held a fascination that propelled the human efforts to explore its surface.

Robot Emissaries and International Gamesmanship

The Moon had been a target for Sergei Korolev and the Soviet space program from before the advent of the space age. Khrushchev supported space exploits because of the propaganda value they held. He, too, recognized the special place the Moon held in the public's perception. Khrushchev approved the use of a modified R-7 rocket, as well as a new and more powerful upper stage, to launch probes to the Moon. Launched under the strictest of secrecy in 1958, the first three probes all failed. The first, named *Ye-1*, launched on September 23, 1958, consisted of a pressurized sphere similar to *Sputnik 1* and carried five scientific instruments that could measure the gas component of interplanetary matter, meteoric particles, cosmic rays, the magnetic fields of the Moon and Earth, and heavy nuclei in primary cosmic radiation. During launch the booster became uncontrollable and disintegrated after 93 seconds. A second probe failed for the same reason on October 11, at 104 seconds into the flight. A third attempt on December 4 lasted 245 seconds before failing. The Soviet government never divulged these launch attempts.

Korolev believed that the failures occurred because he had been forced to pursue the program without adequate funding and to accelerate the launch attempts before sufficient testing was possible. He knew that infighting in Moscow had contributed to the funding problem. When called to account for these failures, Korolev quipped, "Do you think that only American rockets explode?" Unlike the secret Soviet program, failures of the Pioneer rocket were constantly reported and criticized by U.S. media and politicos.

After the Soviet Union's success with the Luna probes, in March 1959, *Pioneer 5* became the first U.S. spacecraft to fly past the Moon. Between 1958 and 1968 the Soviet Union and the United States sent a total of sixty probes around or to the Moon (Table 1), with the Soviet Union achieving all the major breakthroughs in lunar probe missions (Table 2) as well as robotic exploration (Box 1).

Figure 5. In October 1959 the Soviets sent *Luna 3* to transmit pictures of the far side of the Moon—thereby giving the Soviets an important "first" in lunar exploration. The probe sent back twenty-nine photographs covering 70 percent of the far side. The photographs were degraded by radio interference and low resolution, but many lunar features could be recognized.

Table 1
Probes Sent to the Moon, 1958–1976

	USSR	USA
Number of probes launched	49	29
Successful or partially successful missions	22	15
Number of failures due to launcher malfunctions	18	8
Time from program start to 1st Moon impact	1 year	4 years
Number of attempts before 1st Moon soft landing	10	Success on 1st attempt

Source: Calculated from "Lunar Exploration Timeline," Lunar and Planetary Institute, https://www.lpi.usra.edu/lunar/missions/, accessed 8/28/2018.

TABLE 2

KEY FIRSTS IN LUNAR EXPLORATION, 1958–1967

Achievement	Date		
	Soviet Union	United States	U.S. lag
First Moon probe launch	Sep. 23, 1958	Aug. 17, 1958	37 days in advance
First launch of probe near the Moon	Jan. 4, 1959 (*Luna 1*)	Mar. 5, 1959 (*Pioneer 4*)	2 months
First Moon probe landing	Sep. 14, 1959 (*Luna 2*)	Apr. 26, 1962 (*Ranger 4*)	31 months
First photo of dark side of Moon	Oct. 7, 1959 (*Luna 3*)	Aug. 23, 1966 (*Lunar Orbiter 1*)	70 months
First Moon soft landing	Feb. 3, 1966 (*Luna 9*)	June 2, 1966 (*Surveyor 1*)	4 months
First Moon orbiter	Apr. 3, 1966 (*Luna 10*)	Aug. 12, 1966 (*Lunar Orbiter 1*)	4 months
First Moon soil data	Apr. 3, 1966 (*Luna 10*)	Apr. 19, 1967 (*Surveyor 3*)	12 months

Source: Roger Launius, Colin Fries, and Abe Gibson, "Defining Events in NASA History, 1958–2006," https://history.nasa.gov/Defining-chron.htm, accessed 8/28/2018.

BOX 1: KEY FIRSTS IN ROBOTIC EXPLORATION, 1957–1960

First Earth-circling satellite launched into orbit, *Sputnik 1* (USSR), October 4, 1957.

First living animal in orbit, the dog Laika launched on *Sputnik 2*, November 3, 1957.

First human-made object to escape Earth's gravity and to be placed in orbit around the Sun, *Luna 1* (USSR), January 2, 1959.

First clear images of the Moon's surface, from *Luna 2*, September 12, 1959.

First pictures of the far side of the Moon, taken by *Luna 3*, October 7, 1959.

First return of living creatures from orbital flight, two dogs aboard *Sputnik 5*, August 6, 1960.

The United States did not fare as well in early space-exploration efforts. It tried to match the Soviet successes as best it could, but all the early launches failed. In December 1959, after the failure of the first lunar probes, the army's Jet Propulsion Laboratory started the Ranger project, partly in response to the public relations mess the earlier failures had created. Little improved after NASA's creation in 1958. For example, on August 30, 1961, after JPL had been reassigned to NASA, the first *Ranger* reached space but the launch vehicle placed it in the wrong orbit. Two more attempts in 1961 failed, as did two in 1962. NASA then reorganized the Ranger project and did not attempt another launch until 1964. By that time President John F. Kennedy had pledged that the United States would land Americans on the Moon, and the program had been restructured to aid in learning more about the Moon itself, to ensure that the astronauts would survive. The scientists needed to know the composition and geography of the Moon, and the nature of the lunar surface. Was it solid enough to support a lander, or was it composed of dust that would swallow up the spacecraft? Would communications systems work on the Moon? Would other factors—geology, geography, radiation, and so on—affect the astronauts?

To answer these questions three distinct satellite research programs emerged. The restructured Ranger program was the first of these. NASA's engineers eliminated all scientific instruments from Ranger crafts, leaving only a television camera. Ranger's sole remaining objective was to go out in a blaze of glory as it crashed into the Moon while taking high-resolution pictures. On July 31, 1964, the seventh *Ranger* transmitted 4,316 beautiful, high-resolution pictures of the lunar Sea of Clouds. The eighth and ninth *Rangers* also worked well. Even so, the United States seemingly continued to lag the Soviet Union in the Cold War battlefield of technological competition in space.

The second project was Lunar Orbiter, an effort approved in

1960 to place probes in orbit around the Moon. This project, originally not intended to support Apollo, was reconfigured in 1962 and 1963 to further the Kennedy mandate more specifically by mapping the surface of the Moon and photographing about 95 percent of its surface—more than 14 million square miles—to aid in the selection of landing sites for Apollo astronauts. In addition to a powerful camera capable of sending photographs to Earth tracking stations, it carried three scientific experiments—selenodesy (the lunar equivalent of geodesy), meteoroid detection, and radiation measurement. While the returns from these instruments interested scientists in and of themselves, they were critical to Apollo. NASA launched five *Lunar Orbiter* satellites between August 10, 1966, and August 1, 1967, each successfully achieving its objectives. At the completion of the third mission, moreover, the Apollo planners announced that they had sufficient data to press on with an astronaut landing and could use the last two missions for other activities.

Lunar Orbiter's bounty of images came from a unique onboard photographic system. Instead of sending television pictures back to Earth as electrical signals, *Lunar Orbiter* took actual photographs, developed them on board, then scanned them via a special photoelectric system. Because of this self-contained "darkroom" capacity, scientists referred to *Lunar Orbiter* as the "flying drugstore." Because of concerns that radiation in space might fog photographic film, the system used a slow-speed film. To prevent blurring, the spacecraft compensated for the relatively long exposure times by constantly adjusting its optics and in some cases moving slightly. The resulting images had exceptional quality and provided resolutions of up to three feet from an altitude of thirty nautical miles. For example, the *Lunar Orbiter* spacecraft took an iconic image of Copernicus Crater from the perspective one would get by flying over it at low altitude. After depleting the spacecraft's film supplies, flight controllers purposely crashed all five *Lunar Orbiters* onto the

Moon so that their radio transmitters would not interfere with future spacecraft.

Meantime, the Soviet Union also sent numerous spacecraft to the Moon. Its Luna program, sometimes called Lunik, sent fifteen successful spacecraft to the Moon between 1959 and 1976, many of which achieved firsts in the space race. While the Soviets had many failures, not publicly acknowledged at the time, the importance of this program should be acknowledged:

- *Luna 2* mission made the first successful impact upon the lunar surface, the first human-made object to do so (1959)
- *Luna 3* returned the first photographs of the Moon's far side (1959)
- *Luna 9* was the first probe to achieve a soft landing on another body, the Moon (1966)
- *Luna 10* became the first artificial satellite of the Moon (1966)
- *Luna 17* (1969) and *Luna 21* (1973) deployed roving vehicles that roamed the Moon
- *Luna 16* (1970), *Luna 20* (1972), and *Luna 24* (1976) returned samples to the Soviet Union from the lunar surface

Robotic Moon Soft-Landers

Like so many other points of intersection, soft-landing on the Moon with robotic probes also proved a venue for competition for the United States and the Soviet Union in the 1960s. The Soviets won that competition February 3, 1966, by sending *Luna 9* to the Moon's Oceanus Procellarum region, where it became the first spacecraft to soft-land on another planetary body. Equipped only with a camera and communication equipment, *Luna 9* provided the first dramatic panoramic views of the surface of the Moon.

Arriving on the lunar surface, *Luna 9* landed west of craters

Reiner and Marius. *Luna 9* was designed to release a landing pod immediately before striking the surface. A hinged arm reaching forward from the retrorocket was used to detect the surface. The egg-shaped pressurized pod, weighing 250 pounds on Earth, detached and rolled across the surface. After four minutes four spring-loaded "petals" unfolded, and a thirty-inch radio antenna extended. This design ensured that the pod would operate "right side up." Images showed rocks ranging from four to eight inches distributed on the surface. The television camera on *Luna 9* was able to rotate 360 degrees, although imagery was not returned of the entire view because the spacecraft landed at an angle. Between the second and third transmissions from the Moon, *Luna 9*'s inclination changed from 16.5 degrees to 27.5, perhaps because thermal expansion and contraction caused rocks beneath the lander to shift.

Months after the Soviet Union's success with *Luna 9*, the United States succeeded in soft-landing *Surveyor 1* on the Moon, on June 2, 1966. Carrying two cameras, *Surveyor 1* provided multiple images of the surrounding lunar terrain and nearby surface materials. *Surveyor 1* was placed into a direct trajectory to the Moon's surface. During its final descent, it was slowed by a solid-propellant retrorocket. This element was then jettisoned, removing about 60 percent of *Surveyor*'s mass. Three small engines slowed the lander's velocity to about three miles per hour. At that point the rockets were shut off and *Surveyor* coasted to a gentle landing.

Surveyor 1 photographed and studied the soil of a flat area inside a sixty-mile crater north of Flamsteed Crater in southwest Oceanus Procellarum. The television system transmitted 10,338 photos before nightfall on June 14. The spacecraft also acquired data on the radar reflectivity and bearing strength of the lunar surface, as well as spacecraft temperatures for use in the analysis of the lunar surface temperatures. *Surveyor 1*'s mission was terminated by a dramatic drop in battery voltage.

The Soviet Union remained committed to robotic lunar exploration as well. It landed a second soft-lander on the surface with *Luna 13* on December 24, 1966, in the region of Oceanus Procellarum. The same petal encasement as used on *Luna 9* was opened, antennas were erected, and radio transmissions to Earth began four minutes after the landing. Unlike the earlier probe, *Luna 13* carried new instruments mounted at the end of folding five-foot arms. The first experiment tested density of the lunar soil. One end of the arm was slammed into the soil using a small explosive charge, and the seismic waves reflecting back to the surface were measured. The other arms carried radiation density meters, which exposed the lunar soil to gamma rays and measured the reflected energy. On December 25 and 26, the spacecraft television system transmitted panoramas of the nearby lunar landscape at different Sun angles. Each panorama required approximately one hundred minutes to transmit.

After the failure of *Surveyor 2* on September 22, 1966, NASA's *Surveyor 3* successfully soft-landed on the lunar surface on April 17, 1967, and provided imagery and soil analysis. The lander "bounced" more than once on the surface before coming to rest. Footprints from the initial impact were visible from the final landing site. Besides a camera similar to that on *Surveyor 1*, this lander also carried a mechanical scoop that dug several small trenches in the lunar soil. Over the next three weeks the camera returned more than sixty-three hundred images showing the surrounding rocks and the movements of the scoop. Two years after landing, *Surveyor 3* was visited by the *Apollo 12* astronauts. The television camera and other sections were removed and returned to Earth. The camera was later put on display in the Smithsonian Institution's National Air and Space Museum.

Although NASA lost contact with *Surveyor 4* on July 17, 1967, it followed with *Surveyors 5*, *6*, and *7* over the next few months. While on its trajectory to the Moon, *Surveyor 5* experienced serious prob-

Figure 6. Charles Conrad, Jr., *Apollo 12* commander, examines the *Surveyor 3* spacecraft during the second extravehicular activity (EVA 2) in 1969. The lunar module (LM) *Intrepid* is in the right background. This picture was taken by astronaut Alan L. Bean, lunar module pilot. *Apollo 12* landed on the Moon's Ocean of Storms, only six hundred feet from *Surveyor 3*. The television camera and several other components were taken from *Surveyor 3* and brought back to Earth for scientific analysis. *Surveyor 3* soft-landed on the Moon on April 20, 1967.

lems with a helium pressurization system that was necessary for the retrorockets to work. Flight engineers were able to work around the problem, and *Surveyor 5* successfully landed on September 10, 1967. Thousands of images were returned by the television camera. *Surveyor 5* also carried an alpha ray scatterer that measured composition of the lunar soil. *Surveyor 6* landed on November 9, 1967. It carried instruments similar to *Surveyor 5*'s. On November 17, after controllers noted that enough fuel remained for a brief firing of the retrorockets, *Surveyor 6* became the first spacecraft to take off from the lunar surface. When the retrorockets were fired, it performed a "hop," reaching a height of about ten feet and coming to rest about eight feet from its previous position. Both sets of footprints in the lunar soil were plainly visible in images from the television camera. *Surveyor 7* landed on January 10, 1968, north of the crater Tycho. *Surveyor 7* carried both a mechanical arm and an alpha scattering instrument. The arm was needed to move the latter device when it was found to be stuck. Over three weeks, the alpha scattering sensor was lowered and then moved to test composition of soil from the surface and within trenches.

Five of the seven *Surveyor* spacecraft completed their missions, although the vernier rockets on *Surveyor 3* did not shut down at the proper point, causing the 650-pound robot to skip twice across the lunar surface before stopping beneath a small crater rim.

Making the Most of Robotic Lunar Exploration

The Soviet Union continued its effort to explore the Moon well into the 1970s, although the American robotic effort ended with the Surveyor program. During the space race of the 1960s the United States and the Soviet Union competed for space firsts across a broad front of human and robotic missions targeted on the Moon, Venus, and Mars. The United States won the most significant prize by landing a human on the Moon first, demonstrating its capability so thoroughly that the Soviet Union withdrew from the race, disingenuously claiming that it had not been racing the United States at all. Instead, it succeeded in landing the first robotic rovers on the Moon, *Lunokhods 1* and 2.

The first rover, *Lunokhod 1*, was soft-landed on the lunar surface by *Luna 17* on November 17, 1970. Weighing just under one ton, this rover was intended to operate for ninety days while guided from the Soviet mission control center outside Moscow. It had been launched on November 10, and flight controllers undertook two course corrections to bring it into lunar orbit on November 15. After landing, this eight-wheeled rover departed *Luna 17* by means of ramps extending from both sides of the lander. It was the first roving remote-controlled robot to land on another world. *Lunokhod 1* far exceeded its design life, and traveled around the lunar Mare Imbrium (Sea of Rains) for eleven months after landing. It ran only during the two-week-long lunar day, stopping occasionally to recharge its batteries via its solar panels. The rover's operations officially ceased on October 4, 1971, on the fourteenth anniversary of the launch of *Sputnik 1*.

Viewed initially as a scout for a landing site for Soviet cosmonauts, this rover had originated in 1963 and was to carry a beacon that would be used to guide the cosmonauts to the surface. By the time that it flew, however, this mission had been preempted by events. *Lunokhod 1* therefore explored the lunar surface and returned scientific information about the geology and landscape of the Moon, imaging its surface, undertaking laser-ranging experiments from Earth, and measuring magnetic fields. It carried a cone-shaped antenna, a highly directional helical antenna, four television cameras, and an extendable arm to test the lunar soil for soil density. Soviet scientists also included an X-ray spectrometer, an X-ray telescope, cosmic-ray detectors, and a laser-ranging device.

Lunokhod 1 had a unique design; it looked like a bathtub on eight wheels. The inside of the large convex lid served as the solar array, and the tub itself housed the instruments. Using imagery from a large panoramic camera, a five-person team of controllers on Earth sent commands to the rover in real time to control its movement. By the end of its service life *Lunokhod 1* had transmitted more than twenty thousand TV pictures and more than two hundred TV panoramas. It also conducted more than five hundred lunar soil tests.

The exact location of *Lunokhod 1* on the lunar surface is uncertain because laser-ranging experiments have been unable to detect a return signal since the end of the mission. But it traveled more than 10 kilometers (6.2 miles); by comparison, American astronauts during six landing missions traveled on the surface between 10.1 meters (33.2 feet) on *Apollo 11* and 35.4 kilometers (nearly 22 miles) on *Apollo 17*.

Lunokhod 2, launched on the *Luna 21* spacecraft on January 8, 1973, was a virtual twin of its predecessor. After a midcourse correction the day after launch, *Luna 21* entered orbit around the Moon on January 12. Its orbital parameters, the maximum and minimum heights of an orbit as well as its deviation (inclination) from the lunar equator, were 62 by 56 miles at a 60-degree inclination. *Luna 21* soft-landed on the Moon between Mare Serenitatis and the Taurus Mountains and deployed *Lunokhod 2* on January 15, less than three hours after landing. Like the earlier, *Lunokhod 2* weighed a little less than a ton, but it was an improved version of its predecessor, equipped with a third TV camera, an improved eight-wheel traction system, and additional scientific instrumentation. As with *Lunokhod 1*, this rover's primary objectives included imagery of the lunar surface, laser-ranging experiments, solar X-ray analysis, magnetic field measurements, and testing of the properties of lunar surface material.

By the end of its first lunar day, *Lunokhod 2* had already traveled farther than *Lunokhod 1* in its entire operational life. *Lunokhod 2* operated for four months, covered 23 miles of terrain, including hilly upland areas and rilles, and sent back eighty-six panoramic images and more than eighty thousand TV pictures. It also completed several tests of the surface, laser-ranging measurements, and other experiments. On May 9, 1973, the rover rolled into a crater, and dust covered its solar panels, disrupting power to the vehicle. Mission controllers were unable to salvage the rover. On June 3, TASS, the Soviet news agency, announced that the *Lunokhod 2* mission had been terminated. It remains a target for laser-ranging experiments to the present.

After the end of the *Lunokhod 2* mission, Soviet scientists confided that they had received informal help from American scientists working on the Apollo program, who had given them images of the lunar surface near the *Luna 21* landing site at a conference on planetary exploration in Moscow, January 29–February 2, 1973. Those images had been taken as part of the planning for the December 1972 *Apollo 17* lunar-landing mission that took place in the same region. This was after the landing of the spacecraft, but they proved helpful to the controllers in navigating the rover on its mission on the Moon. A third mission, *Lunokhod 3*, was planned for 1977 but never flew because of a lack of launch vehicles and funding. It is now a museum piece at NPO Lavochkin near Moscow.

In addition, the Soviet Union succeeded with three sample return missions from the Moon. After several failed attempts the first mission, *Luna 16*, returned a small sample (101 grams, about 3½ ounces) from Mare Fecunditatis (Sea of Fertility) in September 1970, between the landings of *Apollos 12* and *14*. A second mission, *Luna 20*, entered lunar orbit on February 18, 1972. On February 21, *Luna 20* soft-landed on the Moon in a mountainous area known as the Apollonius Highlands near Mare Fecunditatis. While on the lunar surface, the panoramic television system returned imagery. It also collected lunar samples by means of an extendable drilling apparatus. The ascent stage of *Luna 20* was launched from the lunar surface on February 22, carrying 30 grams of collected lunar samples in a sealed capsule. It landed in the Soviet Union on February 25. The lunar samples were recovered the following day. A third successful sample return mission, *Luna 24*, landed in the area known as Mare Crisium (Sea of Crisis) on August 18, 1976. Like its predecessor, it used a sample arm and drill to collect 170.1 grams

(6 ounces) of lunar samples and deposited them into a collection capsule. The capsule then returned to Earth on August 22, landing in western Siberia. These three Soviet successes salved the open wounds left by the American Apollo program in this most rigorous of all Cold War competitions.

Perceptions

The first race to the Moon may be viewed as the dress rehearsal for the human effort to reach the Moon in the 1960s. As in the human program, the Soviet Union took an early lead in the robotic race. Also as in the human race to the Moon, Soviet early successes did not equate to eventual victory. Indeed, one could make the case that the robotic race was essentially a tie. The successes of the Americans with the Ranger, Lunar Orbiter, and Surveyor programs were very real and keenly felt. It would not be until the successes in the 1970s with its Luna and Lunokhod landers/rovers/sample returns that the USSR recovered a fair measure of respectability in this arena. Of course, the Soviets denied that they had engaged in any race at all. They focused attention on Earth orbital activities, and were successful in launching a succession of space stations.

As the early robotic race proceeded, however, both the Americans and the Soviets prepared for a human effort to reach the Moon. Korolev established a human program to orbit cosmonauts around Earth with the intention of preparing the way for missions to the Moon and planets. The Americans followed suit. The result was an ebb and flow by each of the two nations of activities involving humans aboard spacecraft. The Americans announced a lunar mission first, but the Soviets soon responded, without publicly announcing their intentions, and the race was on.

THREE

Star Voyagers

The experiences of the astronauts and the cosmonauts as space explorers dominated all aspects of the story of the race to the Moon in the 1960s. No one could have predicted the public fascination with the astronauts, from the first unveiling of the Mercury Seven in 1959 through Project Apollo until the present. The astronaut as celebrity, and the effect of that celebrity in American culture, was completely unanticipated. The cosmonauts, less publicly available, still became celebrities in their own right. In both nations the pilots and passengers of spacecraft appeared at a time when each space program desperately needed to inspire public trust in its ability to carry out the nation's goals in space.

Rockets might explode, but the astronauts shined. They seemed to embody the personal qualities in which Americans of that era wanted to believe: bravery, honesty, love of God and country, and family devotion. Similar symbolism prevailed in the Soviet Union. The cosmonaut program had been kept quiet in that communist country until it was publicly unveiled after the success of Yuri

Gagarin's flight in April 1961. The selection of cosmonauts, and the decisions about who flew, when, and in what order was protected as a state secret until the end of the Cold War. The opposing programs evolved an unexpected symmetry.

Meet the Mercury Seven

Just a few days after the establishment of NASA in October 1958, Robert R. Gilruth received approval from NASA Administrator T. Keith Glennan to proceed with what soon became known as Project Mercury, the nation's inaugural human spaceflight program. Concurrent with the decision to move forward with Project Mercury, NASA selected and trained the Mercury astronaut corps. President Dwight D. Eisenhower directed that the astronauts be selected from among the armed services' test pilot force. Although this had not been NASA leadership's first inclination, the decision greatly simplified the selection procedure. The inherent riskiness of spaceflight, and the potential national security implications of the program, pointed toward the use of military personnel. The decision also narrowed and refined the candidate pool, giving NASA a reasonable starting point for selection. Furthermore, it made eminent good sense in that NASA envisioned this astronaut corps first as pilots operating experimental flying machines and only later as scientists. Of course, it was as pilots that the first astronauts became American heroes.

NASA pursued a rigorous process to select the eventual astronauts who became known as the Mercury Seven: record reviews, biomedical tests, psychological profiles, and a host of interviews. Sometimes the would-be astronauts played mind games with their psychologists. Although he had applied during the search for the first Mercury astronauts in 1959, Charles "Pete" Conrad did not achieve his goal at that time. He always believed he failed because he was too flippant when meeting psychologists who were screening

Figure 7. The Mercury Seven astronauts in their iconic silver space suits, 1959. From left to right, they are, back row, Alan Shepard, Virgil "Gus" Grissom, and L. Gordon Cooper; front row, Walter Schirra, Donald "Deke" Slayton, John Glenn, and Scott Carpenter.

the candidates. Conrad's sense of humor became legendary at NASA, and his wit, charisma, and comedy made him the favorite astronaut of many. As Conrad liked to say, "If you can't be good, be colorful." Conrad enjoyed telling the story, which seemed to get better with every rendition, that when a psychologist showed him a blank white card and asked him to describe what he saw, he replied that it was upside down. He was later successful and entered the NASA astronaut corps in September 1962 as a member of the second class brought in to fly during Project Gemini.

Nonetheless, without conclusive results from medical tests, late in March 1959 Robert R. Gilruth's Space Task Group began phase five of the selection, narrowing the candidates to eighteen. Thereafter, final criteria for selecting the candidates reverted to the technical qualifications of the men and the technical requirements of the program, as judged by Charles J. Donlan and his team members. "We looked for real men and valuable experience," said Donlan, and he pressed Gilruth to select the epitome of American masculinity. Gilruth finally decided to select seven. The seven men became heroes in the eyes of the American public almost immediately, in part due to a deal they made with *Life* magazine for exclusive rights to their stories. To most Americans, who knew next to nothing about the organization, the Mercury Seven became the personification of NASA.

Despite the wishes of the NASA leadership, the fame of the astronauts quickly grew out of proportion to their activities. Perhaps it was inevitable that the astronauts were destined for premature adulation, what with the enormous public curiosity about them, the risk they would take in spaceflight, and their exotic training activities. But the power of commercial competition for publicity and the pressure for political prestige in the space race also whetted an insatiable public appetite for this new kind of celebrity. Walter T. Bonney, long a public information officer for the NACA and in 1959 NASA's chief adviser on these matters, foresaw the public and press attention, asked for an enlarged staff, and laid the guidelines for public affairs operations that could maximize the significance of the astronauts as celebrities.

Bonney's foresight proved itself in 1959, only a week after the cherry blossoms bloomed along the tidal basin in Washington, D.C., drenching the city in spectacular spring colors. NASA had chosen to unveil the first Americans to fly in space that April 9. This event made the astronauts public figures. Beforehand, they were a crew-cut, military-minded, mad-monk, thrill-seeking, hard-drinking, woman-chasing, flying-fool gang of daredevils. Now, suddenly, they became heroes of a nation. Excitement bristled in Washington at the prospect of learning who those space travelers might be. Surely they were the best the nation had to offer, modern versions of medieval knights of the Round Table, whose honor and virtue were beyond reproach. Certainly, they carried on their shoulders all the hopes and dreams and best wishes of a nation as they engaged in single combat the ominous specter of communism. The fundamental purpose of Project Mercury was to determine whether humans could survive the rigors of liftoff and orbit in the harsh environment of space. From this perspective, the astronauts were not comparable to earlier explorers who had directed their own exploits. Comparisons between them and Christopher Columbus, Admiral Richard Byrd, and Sir Edmund Hillary left the astronauts standing in the shadows.

At the same time the astronauts, as well as the Soviet cosmonauts, were essentially going off to do personal combat in the Cold War. Each group stood for their separate nations, political systems, and economic approaches against presumed rivals. The United States and the Soviet Union couldn't shoot their ballistic missiles at each other, at least not without ending human existence on this planet, but they could dispatch their space explorers on them and use them as surrogates for outright war.

NASA's makeshift headquarters was abuzz with excitement. Employees had turned the largest room of the second floor, once a ballroom, into a hastily set-up press briefing room, inadequate for the task. Print and electronic media jammed into the room to see the first astronauts. One end of the room sported a stage, complete with curtain, and both NASA officials and the newly chosen astronauts waited behind it for the press conference to begin at 2:00 P.M. The other end had trip hazards of electrical cable strewn about the floor, banks of hot lights mounted to illuminate the stage, and more than a few television cameras that would be carrying the event live and movie cameras recording footage for later use. News photographers gathered at the foot of the stage, and journalists of all stripes occupied seats in the gallery. Seating was insufficient for the media jamming into the ballroom, and NASA employees brought in more chairs and tried to make the journalists as comfortable as possible in the cramped surroundings.

NASA Administrator T. Keith Glennan served as ringmaster for a circuslike press conference to introduce those astronauts. The role did not suit him, and furthermore, he did not comprehend the excitement. But he would play his role. He watched the seven young men chosen at the conclusion of the lengthy selection process take their seats, none of them yet forty years old but each with more than a lifetime's worth of exciting memories. For all of them, he realized, it was the most important event of their lives. But what did it portend, honor and glory or death and contrition? Either seemed likely to Glennan on that bright April afternoon, for NASA's ability to fly people in space seemed somehow distant and fraught with folly despite all the efforts made thus far.

Many of the Mercury Seven astronauts have recorded their recollections of this singular event, and all expressed the same hesitation and dread that Glennan experienced. They also expressed irritation at the huge and unruly audience assembled for the press conference. Alan B. Shepard and Deke Slayton had a brief conversation as they sat down at the table behind the curtain and contemplated the event ahead:

> "Shepard," Deke leaned toward him. "I'm nervous as hell. You ever take part in something like this?"
>
> Alan grinned. "Naw." He raised an eyebrow. "Well, not really. Anyway, I hope it's over in a hurry."
>
> "Uh huh. Me, too," Deke said quickly.

Each of the seven looked at the large Atlas-Mercury rocket and the Mercury spacecraft models set before them.

When the curtains went up, NASA public affairs officer par excellence Walt Bonney announced:

> Ladies and gentlemen, may I have your attention, please. The rules of this briefing are very simple. In about sixty seconds we will give you the announcement that you have been waiting for: the names of the seven volunteers who will become the Mercury astronaut team. Following the distribution of the kit—and this will be done as speedily as possible—those of you who have p.m. deadline problems had better dash for your phones. We will have about a ten- or twelve-minute break during which the gentlemen will be available for picture taking.

As if a dam had broken, a sea of photographers moved forward and popped flashbulbs in the faces of the Mercury Seven astronauts. A buzz in the conference room rose to a roar as this photo shoot proceeded. Some of the journalists bolted for the door with the press kit to file their stories for the evening papers, others ogled the astronauts.

Fifteen minutes later Bonney brought the room to order and asked Keith Glennan to come out and formally introduce the astronauts. Glennan offered a brief welcome and added, "It is my pleasure to introduce to you—and I consider it a very real honor, gentlemen —Malcolm S. Carpenter, Leroy G. Cooper, John H. Glenn, Jr., Virgil I. Grissom, Walter M. Schirra, Jr., Alan B. Shepard, Jr., and Donald K. Slayton . . . the nation's Mercury Astronauts!" These personable pilots faced the audience in civilian dress, and many people forgot that the men were volunteer test subjects and military officers. Rather they were a contingent of mature middle-class Americans, average in build and visage, family men all, college-educated as engineers, possessing excellent health, and professionally committed to flying advanced aircraft.

The reaction was nothing short of an eruption. Applause drowned out the rest of the NASA officials' remarks. Journalists rose to their feet in a standing ovation. Even the photographers crouched at the foot of the stage rose in acclamation of the Mercury Seven. A wave of excitement circulated through the press conference like no one at NASA had ever seen before. What was all the excitement about?

The astronauts asked themselves the same question. Slayton nudged Shepard and whispered in his ear, "They're applauding us like we've already done something, like we were heroes or something." It was clear to all that Project Mercury, the astronauts themselves, and the American space-exploration program were destined to be something extraordinary in the nation's history.

The rest of the press conference was as exuberant as the introduction. At first the newly selected astronauts replied to the press corps's questions with military stiffness, but led by an effervescent and sentimental John Glenn, they soon warmed to the interviews. What really surprised the astronauts, however, was the nature of the questions most often asked. The reporters did not seem to care about their flying experience, although all had been military test pilots, many had combat experience and decorations for valor, and some held aircraft speed and endurance records. The reporters did not seem to care about the details of NASA's plans for Project Mercury, either. They wanted to know about the personal lives of the astronauts. The media wanted to know whether the astronauts believed in God and practiced any religion. They wanted to know whether they were married and the names and ages and genders of their children, they wanted to know what their families thought about space exploration and their roles in it, and they wanted to know about their devotion to their country. God, country, family, and self, and the virtues inherent in each of them, represented the sum total of the reporters' interests.

It was an odd press conference, the reporters probing the character of the pilots. But the reporters' motivation was not to dig up dirt on their subjects, as has long been a media preoccupation, and might easily have been a focus with these men; instead it was just the opposite. The reporters wanted confirmation that these seven men embodied the deepest virtues of the United States. They wanted to demonstrate to their readers that the Mercury Seven strode the Earth as latter-day saviors whose purity and noble deeds would purge this land of the evils of communism by besting the Soviet Union on the world stage. The astronauts did not disappoint.

John Glenn, perhaps intuitively or perhaps through sheer zest and innocence, picked up on the mood of the audience and delivered a ringing sermon on God, country, and family that sent the reporters rushing to their phones for rewrite. He described Wilbur and Orville Wright flipping a coin at Kitty Hawk in 1903 to see who would fly the first airplane and observed how far we had come in only a little more than fifty years. "I think we would be most remiss in our duty," he said, "if we didn't make the fullest use of our talents in volunteering for something that is as important as this is to our country and to the world in general right now. This can mean an awful lot to this country, of course." The other astronauts fell in behind Glenn and eloquently spoke of their sense of duty and destiny as the first Americans to fly in space. Near the end of the meeting, a reporter asked whether they believed they would come back safely from space, and all answered by raising their hands. Glenn raised both of his.

The astronauts emerged as noble champions who would carry the nation's manifest destiny beyond its shores and into space. James Reston of the *New York Times* exulted the astronaut team. He said he felt profoundly moved by the press conference, and that even reading the transcript of it made one's heart beat a little faster and step a little livelier. "What made them so exciting," he wrote, "was not that they said anything new but that they said all the old things with such fierce convictions. . . . They spoke of 'duty' and 'faith' and 'country' like Walt Whitman's pioneers. . . . This is a pretty cynical town, but nobody went away from these young men scoffing at their courage and idealism."

These statements of values seem to have been totally in character for what was a remarkably homogeneous group. They all embraced a traditional lifestyle that reflected the highest ideals of American culture. The astronauts also expressed similar feelings about the role of family members in their lives and the effect of the astronaut career on their spouses and children. Many commentators have remarked on the intertwining of the family and work lives of the astronauts, something also seen in military members and their families as well as among others engaged in professions where lives are placed on the line. Those professions often exacted a toll on all involved in the relationship. In every instance the wives of the astronauts, as well as their children and extended family, experienced the stress of carrying the burden of the nation into space. Many wives commented on how it was not "his" career any longer, but "our" career.

For many Americans the public personas of the wives of the astronauts were just as significant as their famous husbands. As a group these women were always "proud," "thrilled," and "happy"—their watchwords. Their private well-being, however, was often something less. They spent their time in the public eye caring for their families and for one another, while supporting the efforts of NASA in reaching for the Moon. Essentially, the space program called for the wives of astronauts to set the standard for the moral and social well-being of the space agency, to serve as a support for group members in both good and difficult times, and to enforce the principles of the organization. That is exactly what the astronauts' wives did.

The media, reflecting the desires of the American public, scrutinized the astronauts and their families at every opportunity. The insatiable nature of this desire for intimate details prompted NASA to construct boundaries that both protected the astronauts and reinforced the image of the astronauts and their families as models of American society. NASA, for obvious reasons, wanted to portray an image of happily married astronauts, with no hint of extramarital scandal or divorce. Gordon Cooper, one of the Mercury Seven, recalled that public image was important to some inside NASA because "marital unhappiness could lead to a pilot making a wrong decision that might cost lives—his own and others'." That might have been part of it, but the agency's leadership certainly wanted to preserve the image of the astronaut as clean-cut, all-American boy.

The astronauts humanized this endeavor and created the myth of the virtuous, no-nonsense, able professional. In some respects it was a natural occurrence. The Mercury Seven were, in essence, each of us. None was either aristocratic in bearing or elitist in sentiment. They came from everywhere in the nation, excelled in the public schools, trained at their local state universities, served their country in war and peace, married and tried to make lives for themselves and their families, and ultimately rose to their places on the basis of merit. They represented the best we had to offer, and most important they expressed at every opportunity the virtues ensconced in the democratic principles of the republic.

The Cosmonaut as Mirror Image

On the Soviet side the cosmonauts were less public, at least until they made their flights, but equally attractive. In September 1959, chief designer Sergei Korolev established a cosmonaut selection commission under the Scientific Research Institute of the Soviet Air Force. Since American astronauts had just begun their training in the full glare of the world's media, the Soviets, who lacked a focused central organizing body like NASA, were nevertheless able to borrow and adapt much of what they saw of NASA's selection and training methods for their own program. Their training emphasized physical fitness, with all of the first selectees in their late twenties; each of them was a fighter pilot, but not necessarily a test pilot. Korolev developed close relations with his first cosmonauts.

Korolev drew from an initial pool of about three thousand military pilots who had experience flying high-performance aircraft. This group then underwent reviews of their medical and other records, a battery of medical tests, and a range of physical and psychological stress testing. This whittled the group down to fifteen pilots, collectively referred to as Air Force Group One. The group then reported for assignment in March 1960 to a newly formed Tsentr Podgotovka Kosmonavti (TsPK)—the Cosmonaut Training Center—located just outside of Moscow. TsPK later evolved into the Star City facility that is still used for Russia's cosmonaut training today. Of the fifteen who were trained, eleven eventually made at least one spaceflight.

Among the eleven were some of the most famous of all the cosmonauts: Yuri Gagarin, Andrian Nikolayev, Pavel Popovich, Gher-

Figure 8. Most of the original 1960 group of cosmonauts is shown in a photo from May 1961 at the seaside port of Sochi. Sitting in front from left to right: Pavel Popovich, Viktor Gorbatko, Yevgeny Khrunov, Yuri Gagarin, chief designer Sergei Korolev, his wife Nina Koroleva with Popovich's daughter Natasha, Cosmonaut Training Center director Yevgeny Karpov, parachute trainer Nikolay Nikitin, and physician Yevgeny Fedorov. Standing in the second row: Aleksei Leonov, Andrian Nikolayev, Mars Rafikov, Dmitri Zaykin, Boris Volynov, Gherman Titov, Grigori Nelyubov, Valery Bykovsky, and Georgy Shonin. In back: Valentin Filatyev, Ivan Anikeyev, and Pavel Belyayev.

man Titov. These, along with Anatoli Kartashov, who did not fly, and Grigori Grigoryevich Nelyubov, who was dismissed as a cosmonaut in 1963 for disorderly conduct while drunk, were known as the Sochi Six for the location where a famous picture of them was taken. After his dismissal, Nelyubov was airbrushed out of the picture, something not admitted for years by the USSR. Regardless, this group emerged as the elite of the cosmonaut trainees and soon took on greater responsibilities than the rest of the candidates. All the cosmonaut selectees were aware that they were competing for the honor of being the first to fly in space, and by early 1961 Gagarin was effectively viewed as the front-runner, both by Korolev and by the other cosmonauts. Korolev, however, kept his first choice secret so that everyone else would keep working hard. Korolev refused to announce, even internally, who would take on the responsibility of becoming the first cosmonaut to fly aboard the Soviet spacecraft *Vostok 1*—a small vehicle just 7½ feet in diameter, weighing less than three tons, with a ballistic reentry acceleration of eight times the force of Earth's gravity, about 256 feet per second squared—almost until the point of its first launch. Despite his training, Titov reported experiencing disorientation, extreme fatigue, dizziness, and nausea. Titov even found it difficult to differentiate between the Earth and space. This forced a reconsideration of cosmonaut selection and training processes.

In 1962, Korolev selected another group of cosmonaut candidates, their physical capabilities more rigorously assessed. This group also included the first women cosmonauts, and the next year the first scientist-cosmonauts were added to the team. Further additions came in 1964, 1965, and 1966.

As TsPK, outside of Moscow, evolved into Star City, Korolev ensconced the cosmonauts and much of the planning apparatus for human spaceflight there. Highly secretive, Star City did not even appear on maps. It has been the heart of the Russian human space program from its creation until the present. The central goal of the cosmonauts at Star City in the 1960s says much about the priorities of the space race. According to Georgy Grechko, one of the early cosmonauts who trained with Yuri Gagarin: "We thought of nothing else but doing everything before the Americans; it was part of being a good communist."

Like their American counterparts, the cosmonauts were public relations symbols, both within and beyond the Soviet Union. On April 14, 1961, two days after Yuri Gagarin became the first human in space, the Soviet Union held a gigantic ceremony in Red Square in Moscow to honor its first cosmonaut. Gagarin had been chosen for cosmonaut training in 1959 and had undergone a series of increasingly rigorous physical and mental exercises to prepare for his flight. Selected from among several other contenders for the first spaceflight, Gagarin represented the Soviet ideal of the worker who rose through the ranks solely on the basis of merit. His handsome appearance, thoughtful intellectuality, and boyish charm made him an attractive figure on the world stage. The importance of these attributes was not lost on Soviet Premier Nikita Khrushchev and other senior Soviet leaders. The success of the *Vostok 1* flight made the gregarious Gagarin a global hero, and he was an effective spokesman for the Soviet Union on the world stage.

Khrushchev recognized Gagarin's winning personality and sent him on a goodwill tour. Gagarin's spaceflight energized the Soviet leadership to invest more money in space exploration during the years that followed, in part because of the international prestige that the nation gained for its spectacular missions. Subsequently the Soviets were the first to launch a woman, Valentina Tereshkova, into orbit; the first to launch two- and three-person crews; and the first to execute an extravehicular activity, or spacewalk. It was a major loss to the Soviet program when Gagarin died in a plane crash while on a training mission for the Soviet air force on March 27, 1968.

Training the First Space Explorers

Both the Soviet cosmonauts and the American astronauts undertook intense training to prepare for their space-exploration missions, with many exotic devices, complex procedures, and physical endurance tests. Throughout their respective programs, astronauts and cosmonauts underwent medical tests before, during, and after their flights, physical training, procedure training, pilot familiarization training, and plenty of experiment practice. They would simulate all phases of their missions to assure mission planners that they could accomplish their tasks without error, almost by rote.

Those early space explorers were keenly aware that their simple spacecraft allowed for humans to venture into space only for short periods of time. They also knew they had to quickly learn how best to cope with and overcome any technological hurdles they might encounter. Life scientists worked closely with their respective groups of pilots to ensure both survival and success in the mission, and engineers worked to help the space travelers to master their spacecraft's controls and systems. They even had to undertake earthly survival training in case they landed in a remote location and needed to sustain themselves while awaiting rescue.

First and foremost, astronauts and cosmonauts had basic human physiological needs that had to be satisfied in orbit. Scientists, flight surgeons, life scientists, and the fliers themselves worked together to ensure the survival of the human body in the extreme environment of space. They strove to encapsulate the body in protective garments, pressure suits, and space suits of various types, and to attend to the needs of those venturing into space to ensure their survival. Mostly they have been successful, but there have been some notable losses of life.

Many pursued knowledge that would help the body survive. For example, in New Mexico, John Paul Stapp and others tested the limits of the human body in surviving the forces of acceleration by riding a rocket sled simulating what the astronaut would experience in flight. Cosmonauts and astronauts tested their capabilities by riding centrifuges to more than ten *g*s—ten times the force of gravity. Balance and stability were approximated through full-motion simulation, and constant training made actions necessary to survival in space second nature.

Typically, astronaut and cosmonaut training could be split into three stages: individual general training, group training, and specific training for missions. The space explorers learned both technical and management skills focused on understanding, operating, and, if necessary, repairing their respective spacecrafts. Prospective cosmonauts and astronauts developed their own specializations and served as liaisons to the technical teams working on building the space vehicles. For example, John Glenn had responsibility for representing the astronauts in human factors, the manner in which astronauts related to the vehicles they flew. Finally, as a flight neared its launch date, its crew would rehearse and rehearse the details of the mission, often in a full-size mockup of the spacecraft. Usually crew members would become so proficient that they could perform every task automatically without much conscious thinking about it.

The success of such training was obvious throughout the early human space missions. Astronauts and cosmonauts regularly faced glitches and equipment failures, but in each case they were successful in overcoming them.

The First Humans into Space

It came as a shock to the system for many in the United States when Yuri Gagarin became the first human to orbit Earth, on April 12, 1961. The flight of *Vostok 1*, engineered by Sergei Korolev, represented one of the great success stories of the first few years of the space age. The launch of this first Soviet human orbital mission proved enormously important both for Gagarin and for the Soviet Union. *Vostok 1* was a three-ton ball-shaped capsule, with an attached two-ton equipment module containing (among other things) retrorockets. The two parts to the Vostok spacecraft included a sphere in which the pilot sat and a cone-shaped service module that separated as the capsule entered Earth's atmosphere.

It was lofted into orbit atop a modified R-7 ICBM rocket. Gagarin's 108-minute flight had been the direct result of Cold War competition between the United States and the Soviet Union to be the first to send a human into space, as a demonstration of technological superiority before the world. With Gagarin's success, the

United States lost that challenge and the Soviet Union was recognized worldwide as a technological and scientific superpower.

It was not a given that Gagarin would be the first to fly, or that he would be successful. There had been many earlier failures, and during the Cold War the Soviet Union hid its space accidents. For example, Soviet officials denied an explosion during launch of a Soviet R-16 ballistic missile at Baikonur Cosmodrome on October 24, 1960, in which Chief Marshal of Artillery Mitrofan Ivanovich Nedelin and as many as 125 other people died. Not until the collapse of the Soviet Union did the Russians confirm this accident, though accounts of it had circulated among Soviet watchers for decades. The chief designer of the R-16 that exploded, Mikhail Yangel, survived, having stepped away from the test to smoke a cigarette behind a bunker. In the USSR's hyperaggressive military environment, Yangel had intended to challenge Korolev's control of the human spaceflight program, but after this accident his stock fell sharply in the Kremlin. Rumors of additional failures in early attempts to launch a cosmonaut into space abounded during the Cold War, with speculation that as many as two cosmonaut deaths had taken place before Yuri Gagarin's successful flight. There is no evidence to support that theory, even though many historians combed Soviet archival material and memoirs of Russian space pioneers after the end of the Cold War to learn the truth.

Following Nedelin's failed R-16 launch attempt, Korolev remained in firm control, and proceeded with the Vostok program, which employed a surplus ballistic missile to lift a capsule to orbit, where it could sustain a human occupant for a day or more. Moreover, a rudimentary solid-fuel rocket engine propelled the spacecraft out of orbit at the end of the flight, an ablative heat shield dissipated the heat produced as the capsule returned through Earth's atmosphere, and a parachute system broke its descent as it landed on the Russian Steppes. On the morning of Yuri Gagarin's flight,

Figure 9. Yuri Gagarin is in the bus on the way to the launch pad on the morning of April 12, 1961, about to make the world's first human flight into space. Behind him, seated, is his backup, Gherman Titov. Standing are cosmonauts Grigori Nelyubov and Andrian Nikolayev.

from a remote and secluded area known as Tyuratam, both he and his backup, Titov, suited up. Korolev was also present as chief engineer. The tension of those working the mission proved excruciating, according to engineers who recollected it many years later. Gagarin was surprisingly calm and confident. He relieved himself on the side of the carrier before entering the capsule at the top of the rocket; that became a ritual engaged in by all cosmonauts ever after.

"Here we go!" Gagarin said as *Vostok 1* lifted off. He was the first person to see Earth from a height of more than 100 miles, and he watched the continents, islands, and rivers slide by as he orbited at 17,500 miles per hour. When he tried to write up what he was seeing and doing in orbit, his pencil floated away. By any standard the flight was risky. The unknowns were very real. They included the safety and reliability of the technology, the still forming processes and procedures, and the ability of any human to withstand the forces of launch, weightlessness, reentry, and landing.

Because of concerns that Gagarin might be incapacitated by the stress of the flight, *Vostok 1* was controlled from the ground. The launch was successful, and after slightly less than one orbit the Vostok capsule reentered Earth's atmosphere. As intended, Gagarin safely ejected from the capsule at 4.4 miles and parachuted to a landing near the capsule. The success of the mission and the subsequent elevation of Gagarin to heroic stature represented a high point for the USSR. Gagarin's accidental death in a military jet crash in 1968 was a blow to the Soviet psyche thereafter.

Despite its outward success, Gagarin's *Vostok 1* flight had several serious problems. Since the end of the Cold War, increasing access of information has confirmed what some analysts believed all along: that Gagarin's flight had nearly been a disaster when the capsule spun dangerously out of control while beginning the reentry sequence. Gagarin told officials during a postflight debriefing, "As soon as the braking rocket shut off, there was a sharp jolt, and the craft began to rotate around its axis at a very high velocity." It spun uncontrollably as the equipment module failed to separate from the cosmonaut's capsule. After ten minutes the spacecraft stabilized somewhat from its dizzying spin. Nevertheless, Gagarin ejected from the capsule and parachuted safely to Earth. The first people to see him were a mother and her child, who stared bewildered at this man who had just fallen out of the sky. "I am one of yours, a Soviet," Gagarin reassured them. "I've come from outer space."

The United States, already shaken by Soviet firsts in space beginning with *Sputnik 1* in 1957, could only react to the Soviet accomplishment and would not match it until John Glenn's flight ten months later. By the fall of 1961 a "space race" between the two nations was in full force.

The Soviet Union guarded the secret of Gagarin's parachuting from the capsule for many years. The Fédération Aéronautique Internationale (FAI), the official body recording aerospace records, required that for a record to be awarded the pilot must land with the vehicle. To ensure the Soviet record, the Soviet Union perpetuated the lie that Gagarin had landed with *Vostok 1*, not admitting the truth until 1971. To claim circumnavigation of Earth, the Soviets also lied about the location of the launch and landing sites; in reality the orbit was not quite the entire circumference of Earth.

The Soviets were also the first to undertake a rendezvous in space. On August 12, 1962, *Vostok 3* and *Vostok 4* reached orbits that passed within several miles of each other but did not have the actual maneuvering capability to perform a true orbital rendezvous. At closest approach the cosmonauts were able to see the glint of sunlight off the distant spacecraft. The same was true of *Vostok 5* and *Vostok 6*, launched June 16, 1963, to repeat the experiment. As Vasily Mishin, Korolev's deputy and a longtime senior official in the Soviet space program wrote in his memoirs:

> The group flight . . . well, a day after the launch, the first craft was over Baikonur. If the second craft were launched now with great precision, then they would turn out to be next to each other in space. And that's what was done. . . . The craft turned out to be 5 kilometers [three miles] from each other! Well, since, with all of the secrecy, we didn't tell the whole truth, the Western experts, who hadn't figured it out, thought that our Vostok was already equipped with orbital approach equipment. As they say, a sleight of hand isn't any kind of fraud. It was more like our competitors deceived themselves all by their lonesome. Of course, we didn't shatter their illusions.

The success of the Soviet Union in space in the earliest years of the space age catapulted it into an unprecedented global limelight. In a closed society such as the Soviet Union, these successful flights collectively signaled a capability that impressed the world's population.

By the first years of the 1960s the Soviet space program had come to resemble, more than anyone appreciated at the time, the personality of Sergei Korolev: cautious, pragmatic, intensely driven, and firmly dedicated to the systematic exploration of space. His commitment to safety prompted him to emphasize technology that could be enhanced and built on rather than used and then discarded, an approach that quickly became a hallmark of Soviet/Russian space operations to the present. Only Voskhod, forced on Korolev by Khrushchev as a prestige program, represented a different tactic; there three cosmonaut seats were substituted for the ejection system so that the Soviet Union could beat the United States in launching a three-person crew into space.

The United States responded after Gagarin's flight with the beginning flights of the Mercury program, but it was only a salve on an open wound when Alan B. Shepard rode a Redstone booster in the *Freedom* 7 Mercury spacecraft to become the first American in space during a fifteen-minute suborbital flight on May 5, 1961. A second suborbital flight, launched on July 21, 1961, had unique problems. Upon landing, the capsule's hatch blew off prematurely from the Mercury capsule, *Liberty Bell* 7, and sank into the Atlantic Ocean before it could be recovered. In the process the astronaut, Virgil I. "Gus" Grissom, nearly drowned before being hoisted to safety in a helicopter. Not until July 20, 1999, did Oceaneering International, Inc., recover *Liberty Bell* 7 from the floor of the Atlantic Ocean. Curt Newport, financed by the Discovery Channel, brought the spacecraft up from a depth of nearly fifteen thousand feet, three hundred nautical miles east-southeast of Kennedy Space Center. Though less pathbreaking that Gagarin's flight, those two suborbital missions proved valuable for NASA technicians, who found ways to solve or work around literally thousands of obstacles to successful spaceflight.

Comparisons between the Soviet and American flights were inevitable after these first flights (Box 2). Gagarin had orbited Earth; Shepard had been more like a cannonball shot from a gun than a

Box 2: Key Firsts in Human Exploration, 1958–1975

First human in space, cosmonaut Yuri Gagarin, flies a one-orbit mission aboard the spacecraft *Vostok 1*, April 12, 1961.

First American in space, astronaut Alan B. Shepard, Jr., flies a suborbital mission aboard the Mercury spacecraft *Freedom* 7, May 5, 1961.

First daylong human spaceflight mission, *Vostok* 2, with cosmonaut Gherman Titov aboard, August 6, 1961.

First American in orbit, John H. Glenn, Jr., makes three orbits aboard the Mercury spacecraft *Friendship* 7, February 20, 1962.

First long-duration spaceflight, cosmonaut Andrian Nikolayev spending four days in space aboard *Vostok 3*, August 11–15, 1962.

First double (rendezvous) flight, *Vostok 3*, with cosmonaut Nikolayev aboard and *Vostok 4*, with cosmonaut Pavel Popovich aboard, August 12, 1962.

Cosmonaut Valery Bykovsky sets a record aboard *Vostok 5* by orbiting Earth eighty-one times, June 14–18, 1963.

First woman in space, cosmonaut Valentina Tereshkova, flies forty-eight orbits aboard *Vostok 6*, June 16–19, 1963.

First multiperson mission into space, *Voskhod 1* carrying cosmonauts Vladimir Komarov, Boris Yegorov, and Konstantin Feoktistov, October 1964.

First spacewalk or extravehicular activity, by Alexei Leonov during the *Voskhod 2* mission, March 1965.

First American spacewalk, by astronaut Edward White during *Gemini 4*, June 3, 1965.

First docking of two spacecraft in Earth orbit, *Gemini 8* with an Agena target vehicle, March 16, 1966.

First human circumlunar flight, *Apollo 8*, December 24–25, 1968.

First human landing on the Moon, *Apollo 11*, July 20, 1969.

First human Moon landing in which the lunar rover allows astronauts to journey far from the touchdown site, *Apollo 15*, July 26–August 7, 1971.

First rendezvous and docking in space of American and Soviet spacecraft, the Apollo-Soyuz Test Project, July 15–24, 1975.

space venturer. Gagarin's Vostok spacecraft had a mass of more than five tons; *Freedom* 7 had a mass of a little more than a ton. Gagarin had been weightless for eighty-nine minutes, Shepard for only five. "Even though the United States is still the strongest military power and leads in many aspects of the space race," wrote journalist Hanson W. Baldwin in the *New York Times* not long after Gagarin's flight, "the world—impressed by the spectacular Soviet firsts—believes we lag militarily and technologically."

As these issues were being resolved, NASA engineers began final preparations for the orbital aspects of Project Mercury. In this phase NASA planned to use a Mercury capsule capable of supporting a human in space not just for minutes, but eventually for as long as three days. As a launch vehicle for this Mercury capsule, NASA used the more powerful Atlas rocket instead of the Redstone. But the decision was not without controversy. Apart from the technical difficulties to be overcome in mating the rocket to the Mercury capsule, the biggest complication was a debate among NASA engineers over the appropriateness of the Atlas for human spaceflight. It was an exceptionally lightweight rocket, made of aluminum pressurized to keep it from crumpling in gravity. Wernher von Braun's rocket team distrusted such a revolutionary design, but was overruled by Robert Gilruth and his engineers at the Space Task Group. Fortunately, the rocket proved more than up to the task of launching the first Americans into orbit.

Most of the concerns had been resolved by the time of the first successful orbital flight of an unoccupied Mercury-Atlas combination in September 1961. On November 29 the final test flight took place, this time with the chimpanzee Enos occupying the capsule for a two-orbit ride before being successfully recovered in an ocean landing. Not until February 20, 1962, however, could NASA accomplish an orbital flight with an astronaut. On that date John Glenn became the first American to circle Earth, making three orbits in the *Friendship* 7 Mercury spacecraft. The flight was not without problems, however; Glenn flew parts of the last two orbits manually because of an autopilot failure, and he left his retrorocket pack (which normally would be jettisoned) attached to the capsule during reentry because a warning light indicated a loose heat shield.

Glenn's flight provided a healthy boost in national pride, compensating for at least some of the earlier Soviet successes. The public, more than celebrating the technological success, embraced Glenn as a personification of heroism and dignity. Hundreds of requests for personal appearances by Glenn poured into NASA headquarters, and NASA learned much about the power of the astronauts to sway public opinion. The agency leadership allowed Glenn to speak at some events, but more often it substituted other astronauts, and declined many invitations altogether. Among other engagements, Glenn did address a joint session of Congress and participated in several ticker-tape parades around the country. NASA thereby discovered a powerful public relations tool that it has employed ever since.

Three more successful Mercury flights took place during 1962 and 1963. Scott Carpenter made three orbits on May 20, 1962, and on October 3, 1962, Walter Schirra flew six orbits. The capstone of Project Mercury was the May 15–16, 1963, flight of Gordon Cooper, who circled Earth twenty-two times in thirty-four hours. The program accomplished its purpose: to orbit successfully a human in space, explore aspects of tracking and control, and learn about microgravity and other biomedical issues associated with spaceflight. But when the Mercury program ended in 1963, the United States had still not caught up to the Soviet Union in world opinion; a majority still believed that the Americans trailed the Russians in space accomplishments.

The First Woman in Space

For the first decade after *Sputnik* the Soviet Union's space program seemed to succeed in almost every aspect of space exploration it attempted. The Soviets scored another success with the flight of Valentina Tereshkova, who became the first woman in space on June 16, 1963. Tereshkova joined the cosmonaut corps on February 16, 1962, after Sergei Korolev persuaded Soviet premier Nikita Khrushchev to approve a plan to put a woman in space. Unlike the male cosmonauts, who had all been experienced pilots, Tereshkova and the other four women shortlisted for Vostok missions—Valentina Ponomaryova, Tatyana Kuznetsova, Irina Solovyova, and Zhanna Yorkina—were chosen from more than four hundred parachutists. Tereshkova soon rose to preeminence among the women in her training group, and she gained the nod to fly on *Vostok 5*. She trained hard for months in preparation for her mission, excelling at weightless flights, isolation and centrifuge tests, engineering courses, more than 120 parachute jumps, and pilot training in MiG-15 jet fighters. Korolev considered her a particularly suitable candidate for the mission because of her humble background as a worker in a textile factory in the Soviet Union. Moreover, her father had been a war hero, killed in the Finnish continuation campaign of the Great Patriotic War, as the Soviets termed World War II.

Originally Korolev planned to have two female cosmonauts in orbit simultaneously, Tereshkova in *Vostok 5* and Ponomaryova in *Vostok 6*, but this plan was altered in March 1963. The two spacecraft would still be launched days apart to be in orbit simultaneously, but Valery Bykovsky was the cosmonaut in *Vostok 5*, and Tereshkova was bumped down to *Vostok 6*. Tereshkova watched Bykovsky's launch on June 14 and followed him into orbit two days later.

Like Gherman Titov, Tereshkova experienced nausea during much of the flight, even vomiting, but she successfully completed forty-eight orbits over three days—a greater total of time spent in space than all of the American astronauts combined to that date. She returned to a hero's welcome, and, like Yuri Gagarin before her, she was sent around the world as a goodwill ambassador.

But while Tereshkova's feat seemed to signal a greater equality between men and women, in both the Soviet space program and under the Soviet system more generally, in reality it represented neither of these things. Tereshkova's mission was essentially a publicity stunt to achieve another space first. No other woman flew in space during the space race, although other women trained and were eager to fly. In 1969 the cadre of women cosmonauts was disbanded. The next woman in space, Svetlana Savitskaya, would not get into orbit until August 19, 1982, more than nineteen years after Tereshkova. Savitskaya was followed the next year by the first American woman to fly in space, Sally K. Ride, who orbited Earth aboard the Space Shuttle mission STS-7.

Tereshkova went on to a distinguished career as a Soviet politician, serving in the Supreme Soviet of the Soviet Union between 1966 and 1974 and in the Presidium of the Supreme Soviet from 1974 to 1989. She also served on several delegations on behalf of the USSR. With the end of the Soviet Union, Tereshkova went on serve in the Russian State Duma in 2011 and remained a member through 2017. Although seventy-one years old at the time, she was a torchbearer for the 2008 Summer Olympics when the torch passed through Saint Petersburg en route to Beijing for the start of the Games.

As the Soviets were sending Tereshkova into space, many in the United States also believed women should have the opportunity to become space explorers. Shortly after the public unveiling of the Mercury Seven astronauts in the United States, William "Randy" Lovelace II, one of the life scientists who had been involved in the selection, began investigating whether women could perform in similar selection tests as well as their male counterparts. After meeting

American aviator Geraldyn "Jerrie" Cobb in 1959, Lovelace invited her to take the same tests as the Mercury astronauts, and was astounded by her aptitude. Uncertain whether Cobb might be an anomaly, Lovelace secured private funding from veteran pilot Jackie Cochran, the head of the Women Airforce Service Pilots (WASP) program in World War II, to bring eighteen other experienced female pilots to his clinic for secret testing.

The women arrived alone or in pairs for four days of tests, including experiments on centrifuges to simulate the strains of launch and reentry. All the women were skilled airplane pilots with commercial ratings, and they fared well on the tests. When word of the experiments leaked to the press, the top twelve, along with Jerrie Cobb, were dubbed the "Mercury 13." Some of them believed that their participation in the tests could lead to their becoming NASA astronauts themselves, but NASA had no direct involvement in Lovelace's work. Indeed, when NASA's leadership learned of the experiments, it soon declared that it had no plans to employ women astronauts.

This did not dissuade the American media, which had a field day with the women's outstanding test results. Some of the choice headlines included: "Astrogals Can't Wait for Space," "Spunky Mom Eyes Heavens," and "Why not 'Astronauttes' Also?" There was a great deal of excitement about America's first "lady astronauts," and several of the women were interviewed for television programs and for photo spreads in magazines and newspapers around the United States.

Cobb lobbied extensively behind the scenes for NASA to accept female astronauts and eventually secured a dramatic congressional hearing in July 1962 to explore the possibility of women qualifying for the program. She and fellow pilot Janey Hart, whose husband was the United States senator Philip Hart (D-MI), used Cold War arguments before the House Committee on Science and Astronautics to support their contention that women should fly in space. Sending an American woman into space first, they reasoned, could score a highly visible accomplishment in the space race. NASA officials at this hearing, among them astronaut John Glenn, all insisted that astronauts must be military test pilots and that training women would slow down the astronaut program.

After two days of testimony, the committee concluded that NASA's selection process would not be changing anytime soon. The turning point came when Glenn testified to Congress that there was no place for women in that early program. Later, he would admit that his testimony had reflected both the perspectives of NASA's officials and his personal beliefs at the time. He said he had overcome that thinking in later years. It would be another twenty-one years before an American woman would be sent into space, with Sally Ride serving as a mission specialist on space shuttle flight STS-7 on June 18, 1983. It would be thirty-two years before America sent its first female pilot, Eileen Collins, into space. Shortly before her first launch in 1995, Collins heard the story of the Mercury 13 and invited some of them to Cape Canaveral to witness her launch.

A Meaning for the First Flights

The early American and Soviet human space missions had few objectives other than to determine whether or not astronauts and cosmonauts could survive in the exceptionally harsh environment of space. Both programs learned how difficult it was for fragile *Homo sapiens* to venture into a realm for which it is ill-adapted, resulting in the necessity of designing systems that enable humans both to survive and to complete useful work. As space life scientist Vadim Rygalov from the University of North Dakota remarked in 2008, "Spaceflight is first and foremost about providing the basics of human physiological needs in an environment in which they do not exist." From the most critical—meaning that its absence would cause immediate death—to the least critical, these include such constants available on Earth of atmospheric pressure, breathable oxygen, tolerable temperature, safe drinking water, digestible food, bearable gravitational pull on physical systems, radiation mitigation, and others of a less immediate nature. Every human spaceflight vehicle, every space suit, every subsystem of even the most simple design takes this as its raison d'être because of the extreme hostility of the space environment.

Having confronted the challenges and dangers of being among the first beings from Earth to venture into space, the astronauts and cosmonauts rightly became heroes for their respective nations and the remainder of the world. Both sides in the Cold War recognized that their "star voyagers" could be used for propaganda purposes to sway the peoples of various nations to their side. Both groups proved effective for this purpose, and some, such as Yuri Gagarin and John Glenn, became superstars in the public sphere. Sending astronauts and cosmonauts on world tours, having them address Congress and the Politburo, and reaching broad audiences through television created a public frenzy over the space race. Many people around the world paused as major spaceflight events took place during the 1960s to see how these individuals fared. The drama and excitement of the space race found personal connections to millions through the actions of the cosmonauts and the astronauts.

FOUR

The Decisions to Go to the Moon

The decision of President John F. Kennedy in May 1961 to send Americans to the Moon has left an indelible mark on public perceptions of spaceflight and American culture. The Moon, so wrapped up in the human romance, proved the perfect target for Kennedy's resolve. No other human space mission would have been so innately attractive, so positively viewed in human psyche, so representative of success for a spacefaring nation. That, of course, was the reasoning behind Kennedy's lunar-landing decision and the reason why American society agreed to support it with significant resources—ultimately $24.5 billion—expended over more than a decade. But the lunar-landing decision has taken on a mythical significance as individuals reflect on it from the more than forty years since it took place.

In this chapter I revisit the process of public policy formulation that led to the decision to go to the Moon. In so doing I shall seek to answer several core questions about the lunar-landing decision. First, what processes led to the Apollo decision as it unfolded in

April–May 1961? Second, what was the relationship of this decision to actions in the Soviet Union, and how did the Soviets respond? Not until 1963 did the Soviet Union reach a decision to make a race to the Moon the key element of its space race strategy, even as it denied such a race was under way.

Kennedy and the Early Definition of Space Policy

On May 25, 1961, President Kennedy announced to the nation a goal of sending an American to the Moon before the end of the decade. This decision involved much study and review before it was made public, and tremendous expenditure and effort to make it a reality by 1969. Only the building of the Panama Canal rivaled the Apollo program as the largest nonmilitary technological endeavor ever undertaken by the United States; only the Manhattan Project was comparable in a wartime setting. The human spaceflight imperative was a direct outgrowth of it; Projects Mercury (at least in its latter stages), Gemini, and Apollo were each designed to execute it.

In 1960 JFK, a U.S. senator from Massachusetts between 1953 and 1960, ran for president as the Democratic candidate, with party wheelhorse Lyndon B. Johnson as his running mate. Using the slogan "Let's get this country moving again," Kennedy charged the Republican Eisenhower administration with having done nothing about the myriad social, economic, and international problems that festered in the 1950s. He was especially hard on Eisenhower's record in international relations, taking a Cold Warrior position on a supposed "missile gap" (which turned out to be fictitious): the United States, Kennedy claimed, lagged far behind the Soviet Union in ICBM technology. He also invoked the Cold War rhetoric envisioning a communist effort to take over the world and used as his evidence the 1959 revolution in Cuba, which brought leftist dictator Fidel Castro to power. The Republican candidate, Richard M.

Nixon, who had been Eisenhower's vice president, tried to defend his mentor's record, but when the results were in, Kennedy was elected, with a narrow victory margin of 118,550 out of more than 68 million popular votes cast. His Electoral College victory was more decisive, by a margin of 303–219.

Kennedy as president had little direct interest in the U.S. space program. He was not a visionary enraptured with the romantic image of the last American frontier in space and consumed by the adventure of exploring the unknown. He *was*, on the other hand, a Cold Warrior with a keen sense of realpolitik in foreign affairs, and he worked hard to maintain a balance of power and spheres of influence in American-Soviet relations. The Soviet Union's non-military accomplishments in space forced Kennedy to respond and to serve notice that the United States was every bit as capable in the space arena as the Soviets. Of course, to prove this assertion, Kennedy had to be willing to commit national resources to NASA and the civil space program. The Cold War realities of the time therefore served as the primary vehicle for an expansion of NASA's activities and for the definition of Project Apollo as the premier civil space effort of the nation. Even more significant, from Kennedy's perspective the Cold War necessitated the expansion of the military space program, especially the development of ICBMs and satellite reconnaissance systems.

While Kennedy was preparing to take office, he appointed an ad hoc committee headed by Jerome B. Wiesner of the Massachusetts Institute of Technology to offer suggestions for American efforts in space. Wiesner, who later headed the President's Science Advisory Committee under Kennedy, concluded that the issue of "national prestige" was too great to allow the Soviet Union leadership in space efforts; therefore the United States needed to enter the field in a substantive way. "Space exploration and exploits," Wiesner wrote in a January 10, 1961, report to the president-elect, "have captured the imagination of the peoples of the world. During the next few years the prestige of the United States will in part be determined by the leadership we demonstrate in space activities."

Wiesner also emphasized the importance of practical nonmilitary applications of space technology—communications, mapping, and weather satellites among others—and the necessity of keeping up the effort to exploit space for national security through such technologies as ICBMs and reconnaissance satellites. He tended to deemphasize the human spaceflight initiative for very practical reasons. American launch vehicle technology, he argued, was not well developed, and the potential of placing an astronaut in space before the Soviets was slim. He thought human spaceflight was a high-risk enterprise with a low chance of success. Human spaceflight was also less likely to yield valuable scientific results, and the United States, Wiesner thought, should play to its strength in space science, where important results had already been achieved.

Kennedy accepted only part of what Wiesner recommended. He was committed to conducting a more vigorous space program than had Eisenhower, but he was more interested in human spaceflight than either his predecessor or his science adviser. This was partly because of the drama surrounding Project Mercury and the seven astronauts that NASA was training. Wiesner had cautioned Kennedy about the hyperbole associated with human spaceflight. "Indeed, by having placed the highest national priority on the MERCURY program we have strengthened the popular belief that man in space is the most important aim for our non-military space effort," Wiesner wrote. "The manner in which this program has been publicized in our press has further crystallized such belief." Kennedy, nevertheless, recognized the tremendous public support arising from this program and wanted to ensure that it reflected favorably upon his administration.

But it was a risky enterprise—what if the Soviets were first to send a human into space, what if an astronaut was killed and Mercury was a failure—and the political animal in Kennedy wanted to minimize those risks. The earliest Kennedy pronouncements relative to civil space activity directly addressed these hazards. He offered to cooperate with the Soviet Union, still the only other nation involved in launching satellites, in the exploration of space. In his inaugural address in January 1961 Kennedy spoke directly to Soviet Premier Nikita Khrushchev and asked him to cooperate in exploring "the stars." In JFK's state of the union address ten days later, he asked the Soviet Union "to join us in developing a weather prediction program, in a new communications satellite program, and in preparation for probing the distant planets of Mars and Venus, probes which may someday unlock the deepest secrets of the Universe." Kennedy also publicly called for the peaceful use of space, and the limitation of war in that new environment.

In making these overtures Kennedy accomplished several important political ends. First, he appeared to the world as the statesman, seeking friendly cooperation rather than destructive competition with the Soviet Union, knowing full well that there was little likelihood that Khrushchev would accept his offer. Conversely, the Soviets would appear to be monopolizing space for national, especially military, benefit. Second, Kennedy minimized the goodwill that the Soviet Union enjoyed because of its own success in space vis-à-vis the United States. Finally, if the Soviet Union accepted JFK's call for cooperation, it would tacitly be recognizing the equality of the United States in space activities, something that would also boost American prestige on the world stage.

The Soviet Challenge Renewed

Had the balance of power and prestige between the United States and the Soviet Union remained stable in the spring of 1961, it is quite possible that Kennedy would never have advanced his Moon program, and the direction of American space efforts might have taken a radically different course. Kennedy seemed quite happy to allow NASA to execute Project Mercury at a deliberate pace, working toward the orbiting of an astronaut sometime in the middle of the decade, and to build on the satellite programs that were yielding excellent results both in terms of scientific knowledge and practical application. Jerome Wiesner reflected: "If Kennedy could have opted out of a big space program without hurting the country in his judgment, he would have."

Firm evidence for Kennedy's essential unwillingness to commit to a strong space program came in March 1961 when the NASA administrator, James E. Webb, submitted a request that greatly expanded his agency's budget for fiscal year 1962 in order to permit a Moon landing before the end of the decade. Kennedy's budget director, David E. Bell, objected to this large increase and told Webb that he would have to obtain the president's explicit commitment to make the lunar program a part of the administration's effort "to catch up to the Soviet Union in space performance." At White House meetings March 21–22, 1961, Webb and Bell debated the merits of an aggressive lunar-landing program before Kennedy and Vice President Johnson, but in the end the president was unwilling to obligate the nation to a much bigger and more costly space program. Instead, in good political fashion, he approved a modest increase in the NASA budget to allow for development of the big launch vehicles that would eventually be required to support a Moon landing.

A nonchalant pace might have remained the standard for the U.S. civil space effort had not two important events forced Kennedy to act. The Soviet Union's space effort counted coup on the United States one more time not long after the new president took office. On April 12, 1961, Soviet cosmonaut Yuri Gagarin became the first human in space with a one-orbit mission aboard the *Vostok 1*. The chance to place a human in space ahead of the Soviets had now been lost.

Close in the wake of the Gagarin achievement, the Kennedy administration suffered another devastating blow in the Cold War that contributed to the sense that action had to be taken. Between April 15 and 19, 1961, the Bay of Pigs invasion of Cuba, supported by the administration and designed to overthrow Fidel Castro, failed spectacularly. Executed by anti-Castro Cuban refugees armed and trained by the CIA, the invasion was a debacle almost from the beginning. It was predicated on an assumption that the Cuban people would rise up to welcome the invaders; when that proved to be false, the attack was doomed. American backing of the invasion was a great embarrassment both to Kennedy personally and to his administration. It caused enormous damage to U.S. relations with foreign nations, and made the communist world look all the more invincible.

While the Bay of Pigs invasion was never mentioned explicitly as a reason for stepping up U.S. efforts in space, the international situation certainly played a role as Kennedy scrambled to recover a measure of national dignity. Wiesner reflected, "I don't think anyone can measure it, but I'm sure it [the invasion] had an impact. I think the President felt some pressure to get something else in the foreground." T. Keith Glennan, NASA administrator under Eisenhower, immediately linked the invasion and the Gagarin flight as the seminal events leading to Kennedy's announcement of the Apollo decision. He confided in his diary, "In the aftermath of that [Bay of Pigs] fiasco, and because of the successful orbiting of astronauts by the Soviet Union, it is my opinion that Mr. Kennedy asked for a reevaluation of the nation's space program."

Reevaluating NASA's Priorities

Two days after the Gagarin flight, Kennedy again discussed the possibility of a lunar-landing program with Webb, but the NASA head's conservative estimates of a cost of more than $20 billion for the project was too steep, and Kennedy delayed making a decision. A week later, at the time of the Bay of Pigs invasion, Kennedy called Johnson, who headed the National Aeronautics and Space Council, to the White House to discuss strategy for catching up with the Soviets in space. Johnson agreed to take the matter up with the Space Council and to recommend a course of action. It is likely that one of the programs that Kennedy explicitly asked Johnson to consider was a lunar-landing program, for the next day, April 20, 1961, he followed up with a memorandum to Johnson raising fundamental questions about the project. Kennedy wanted to know whether "we have a chance of beating the Soviets by . . . a trip around the moon, or by a rocket to land on the moon, or by a rocket to go to the moon and back with a man. Is there any other space program which promises dramatic results in which we could win?"

While he waited for the results of Johnson's investigation, this memo made it clear that Kennedy had a pretty good idea of what he wanted to do in space. He confided in a press conference on April 21 that he was leaning toward committing the nation to a large-scale project to land Americans on the Moon. "If we can get to the moon before the Russians, then we should," he said, adding that he had asked his vice president to review options for the space program. This was the first and last time that Kennedy said anything in public about a lunar-landing program until he officially unveiled the plan. It is also clear that Kennedy approached the lunar-landing effort essentially as a response to the competition between the United States and the USSR. For Kennedy the Moon-landing program, conducted in the tense Cold War environment of the early 1960s, was a strategic decision directed toward advancing the far-flung interests of the United States in the international arena. It aimed toward recapturing the prestige that the nation had lost as a result of Soviet successes and U.S. failures. It was, as political scientist John M. Logsdon has suggested, "one of the last major political acts of the Cold War. The Moon Project was chosen to symbolize U.S. strength in the head-to-head global competition with the Soviet Union."

Lyndon Johnson probably understood these circumstances very well, and for the next two weeks his Space Council diligently considered, among other possibilities, an earlier lunar landing than the Soviets could accomplish. As early as April 22, NASA's Hugh L. Dryden had responded to a request for information from the Space Council about a Moon program by writing that there was "a chance for the U.S. to be the first to land a man on the moon and return him to earth if a determined national effort is made." He added that the earliest this feat could be accomplished was 1967, but that to do so would cost about $33 billion, which was $10 billion more than the whole projected NASA budget for the next ten years. A week later Wernher von Braun, director of NASA's George C. Marshall Space Flight Center and head of the big booster program needed for the lunar effort, responded to a similar request for information from Johnson. He told the vice president, "We have a sporting chance of sending a 3 man crew *around the moon* ahead of the Soviets" and "an excellent chance of beating the Soviets to the *first landing of a crew on the moon* (including return capability, of course)." He added that "with an all-out crash program" the United States could achieve a landing by 1967 or 1968.

After gaining these technical opinions, Lyndon Johnson began to poll political leaders for their sense of the propriety of committing the nation to an accelerated space program with Project Apollo as its centerpiece. He brought in Senators Robert Kerr (D-OK) and Styles Bridges (R-NH) and spoke with several U.S. representatives to gauge support for an accelerated space program. Only a few were hesitant, and Robert Kerr worked to allay their concerns. He called on James Webb, who had worked for his business conglomerate during the 1950s, to give him a straight answer about the project's feasibility. Kerr told his congressional colleagues that Webb was enthusiastic about the program, and "that if Jim Webb says we can land a man on the moon and bring him safely home, then it can be done." This endorsement secured considerable political support for the lunar project. Johnson also met with several businessmen and representatives from the aerospace industry and other government agencies to ascertain the consensus of support for a new space initiative. Most of them also expressed support.

General Bernard A. Schriever, commander of the Air Force Systems Command that developed new technologies, expressed the sentiment of many people by suggesting that an accelerated lunar-landing effort "would put a focus on our space program." He believed that it was important for the United States to build international prestige and that the return was more than worth the price to be paid. Secretary of State Dean Rusk, a member of the Space Council, was also a supporter of the initiative because of the Soviet Union's image in the world. He wrote to the Senate Space Committee a little later, "We must respond to their conditions; otherwise we risk a basic misunderstanding on the part of the uncommitted countries, the Soviet Union, and possibly our allies concerning the direction in which power is moving and where long-term advantage lies." It was clear early in these deliberations that Johnson was in favor of an expanded space program in general and a maximum effort to land an astronaut on the Moon. Whenever he heard reservations Johnson used his forceful personality to persuade. "Now," he asked, "would you rather have us be a second-rate nation or should we spend a little money?"

In an interim report to the president on April 28, Johnson concluded, "The U.S. can, if it will, firm up its objectives and employ its resources with a reasonable chance of attaining world leadership in space during this decade"; he recommended committing the nation to a lunar landing. In this exercise Johnson had built, as Kennedy had hoped he would, a strong justification for undertaking Project Apollo, but he had also moved on to develop a greater consensus for the objective among key government and business leaders.

The NASA Position

While NASA's leaders were generally pleased with the course Johnson was recommending—they agreed for the most part with the political reasons for adopting an aggressive lunar-landing program—they wanted to shape it as much as possible to the agency's particular priorities. NASA Administrator James Webb, well known as a skilled political operator who could seize an opportunity, organized a short-term effort to accelerate and expand a long-range NASA master plan for space exploration. A fundamental part of this effort addressed a legitimate concern that the scientific and technological advancements for which NASA had been created not be eclipsed by the political necessities of international rivalries. Webb conveyed the concern of the agency's technical and scientific community to Jerome Wiesner on May 2, 1961, noting that "the most careful consideration must be given to the scientific and technological components of the total program and how to present the picture to the world and to our own nation of a program that has real value and validity and from which solid additions to knowledge can be made, even if every one of the specific so-called 'spectacular' flights or events are done after they have been accomplished by the Russians." He asked that Wiesner help him "make sure that this component of solid, and yet imaginative, total scientific and technological value is built in."

Partly in response to this concern, Johnson asked NASA to provide him with a set of specific recommendations on how a scientifically viable Project Apollo would be accomplished by the end of the decade. What emerged was a comprehensive space policy planning document that had the lunar landing as its centerpiece but that attached several ancillary funding items to enhance the program's scientific value and advance space exploration on a broad front:

1. Spacecraft and boosters for the human flight to the Moon;
2. Scientific satellite probes to survey the Moon;
3. A nuclear rocket;
4. Satellites for global communications;
5. Satellites for weather observation;
6. Scientific projects for Apollo landings.

Johnson accepted these recommendations and passed them to Kennedy, who approved the overall plan.

The last major area of concern was the timing for the Moon landing. The original NASA estimates had given a target date of 1967, but as the project became more crystallized, agency leaders recommended not committing to such a strict deadline. Webb, realizing the problems associated with meeting target dates based on NASA's experience in spaceflight, suggested that the president commit to a landing by the end of the decade, giving the agency another two years to solve any problems that might arise. The White House accepted this proposal.

Decision

President Kennedy unveiled the commitment to execute Project Apollo on May 25, 1961, in a speech on "Urgent National Needs," billed as a second state of the union message. He told Congress that the United States faced extraordinary challenges and needed to respond extraordinarily. In announcing the lunar-landing commitment he said:

> If we are to win the battle that is going on around the world between freedom and tyranny, if we are to win the battle for

> men's minds, the dramatic achievements in space which occurred in recent weeks should have made clear to us all, as did the Sputnik in 1957, the impact of this adventure on the minds of men everywhere who are attempting to make a determination of which road they should take. . . . We go into space because whatever mankind must undertake, free men must fully share.

Then he added: "I believe this Nation should commit itself to achieving the goal, before this decade is out, of landing a man on the moon and returning him safely to earth. No single space project in this period will be more impressive to mankind, or more important for the long-range exploration of space; and none will be so difficult or expensive to accomplish."

Robert Gilruth and Wernher von Braun reacted to the Kennedy decision differently. Gilruth admitted, "Frankly, I was aghast. . . . I wasn't at all sure it could be done." Nonetheless, he accepted the task at hand and made the most of it, though sometimes reluctantly because of concerns for the lives of the astronauts. Von Braun embraced the decision as the culmination of a lifetime of ambitions. Both gave it all they had, sometimes disagreeing about priorities and methods but always keeping the final objective in the distance.

Previously, the American civil space program had been operating at a measured pace, with appropriate long-term goals. In 1959, just over a year after NASA began operation, it prepared a formal long-range plan that announced that its goal in the 1960s "should make feasible the manned exploration of the moon and nearby planets, and this exploration may thus be taken as a long-term goal of NASA activities." The plan called for the "first launching in a program leading to manned circumlunar flight and to a permanent near-earth space station" in the 1965–1967 period. It also called for the first human flight to the Moon at an unspecified time "beyond 1970."

Kennedy threw out the long-range plan by making the Apollo commitment in 1961. In so doing he also overturned the orderly approach to space exploration established during the Eisenhower administration, one that led to the long-range plan and an incremental growth in the budget to about 1 percent of all monies expended by the federal government. Eisenhower had refused to fall prey to public hysteria after the *Sputnik* launches in 1957, and set in place only with some reluctance NASA as an independent Executive Branch agency in 1958. Eisenhower took small steps because he possessed a long-term vision for defeating the Soviet Union in the Cold War without head-to-head competition across a broad spectrum. Indeed, he was committed to achieving without undue cost the development of scientific and technical capability both to gain access to space and to operate therein, but this had to be balanced against a wide range of other concerns.

In the crisis over *Sputnik*, Ike had felt intense pressure from an alliance of diverse interests to establish a cabinet-level federal entity to carry out a visible program of space exploration, something he had always thought unnecessarily expensive and, once created, almost impossible to dismantle. With the creation in 1958 of NASA, an organization with less power and stature than others wanted, Eisenhower was able to deflect the coalition of interests that advocated an exceptionally aggressive space program. In so doing, he thwarted the goal of establishing a large, independent bureaucracy with expensive accelerated programs to race the Soviet Union into space and to accomplish spectacular feats that would impress the world.

Kennedy, however, had a much less refined strategy for how to win the Cold War, and accordingly a greater capacity to view each problem as if he were in a death match. Each confrontation with the Soviet Union took on spectacular proportions and desperate characteristics for Kennedy. For example, had Eisenhower been in office in 1961, it is doubtful that he would have responded to international setbacks with a similar lunar-landing decision. Instead, he probably would have sought to reassure those stampeded by Soviet successes and explain carefully the long-term approach being taken by NASA to explore space. A hint of the Eisenhower approach came in 1962, when he remarked in an article: "Why the great hurry to get to the moon and the planets? We have already demonstrated that in everything except the power of our booster rockets we are leading the world in scientific space exploration. From here on, I think we should proceed in an orderly, scientific way, building one accomplishment on another." He later cautioned that the Moon race "has diverted a disproportionate share of our brain-power and research facilities from equally significant problems, including education and automation."

Kennedy's decision to race the Soviets to the Moon fundamentally altered the space program then under way at NASA, and whether one agrees that this was a positive alteration is very much a matter of perspective. For instance, it placed on hold an integrated space-exploration scenario centered on human movement beyond this planet and involving these basic ingredients accomplished in essentially this order:

1. Earth orbital satellites to learn about the requirements for space technology that must operate in a hostile environment;
2. Earth orbital flights by humans to determine whether it was really possible for humanity to explore and settle other places;
3. Development of a reusable spacecraft for travel to and from Earth orbit, thereby extending the principles of atmospheric flight into space and making space operations routine;

4. Construction of permanently inhabited space stations to observe Earth and from which to launch future expeditions to the Moon and planets;
5. Human exploration of the Moon with the intention of creating Moon bases and eventually permanent colonies;
6. Human expeditions to and eventual colonization of Mars.

Specifically, because of Apollo, NASA lost the rationale for a space station, objective 4, viewed by everyone both then and since as critical for the long-term exploration and development of space.

Instead of building the infrastructure necessary for sustained space exploration, as a space station could have done, JFK committed the nation to a sprint to the Moon as a demonstration of American technological virtuosity, but ultimately it was a demonstration that had little application beyond its propaganda value. Of course, even though the project was not undertaken to advance scientific understanding so much as to resolve Cold War rivalries, one could argue that the scientific return of Apollo was significant. In reality, however, had we found something of interest on the Moon, instead of an aborted space-exploration program, Apollo would have been the vanguard of an armada of spacecraft from Earth. As it was, the belief of most Americans became "been there—done that," and they pushed for decreased funding for NASA and emphases on other projects after the Moon landings. The dreams of sustained human exploration in the solar system were trashed in the perceptions of Apollo as being something only mildly worthwhile for narrow scientific purposes.

The Apollo Decision as a Model of Public Policy Formulation

Analysts have offered four basic approaches to interpreting the Apollo decision-making process in the more than forty years since

Kennedy stood before the American people and declared that we should send astronauts to the Moon. By far the most influential of these interpretations is the conception that Kennedy made a single, rational, pragmatic choice to undertake the U.S. sprint to the Moon as a means of competing with the Soviet Union for international prestige during the height of the Cold War. The president and his advisers therefore undertook an exceptionally deliberate, reasonable, judicious, and logical process to define the problem, analyze the situation, develop a response, and achieve a consensus for action. The timeline progressed from point to point with no cul-de-sacs and few detours from problem definition to sensible decision. It was all so neat and tidy! As such, it has served as a model for public policy formulation.

This argument begins with the assertion that JFK's space policy was a relic of the Cold War struggle between the United States and the Soviet Union, and that it revolved around the question of international prestige. In this view, Apollo was a clear result of competition between the world's two superpowers to win the "minds of men" to a specific economic and political system. In essence, the Apollo program was nothing less than the "moral equivalent of war." It sought to weaken the Soviet Union while enhancing the United States.

There is much to recommend this interpretation, and its study as a model of outstanding policy formulation is appropriate. Its main strength is its insistence that the American effort to land on the Moon served as an enormously effective response to a Cold War crisis with the Soviet Union. At the same time, the most significant problem with this interpretation is its unwavering belief that individuals—and especially groups of individuals, even competing ones—logically assess situations and respond with totally reasonable consensus actions. Since virtually nothing in human existence is done solely on a rational basis, this is a difficult conclusion to accept.

A second interpretation of the Apollo decision suggests that Kennedy's tortured background and aggressive tendencies affected his decision making, causing him to take a more combative approach toward the Soviet Union than required and necessitating his "winning" at whatever challenge came his way. At some level, Kennedy may have even created crisis situations wherein he reaffirmed his quintessential masculinity and enhanced his own dominance over everyone and everything. Most of these analyses depict JFK in an unfavorable light and focus on his tendencies toward the overarching competitiveness, general recklessness, and Machiavellian ambition instilled in all the sons of Joseph P. Kennedy, Sr. These character studies view Kennedy as an individual who had to dominate all, and who unconsciously, or in some cases deliberately, created situations calculated to demonstrate his mastery. His harsh treatment of women, as an ardent and destructive philanderer, demonstrated this fact, as did his competition with all others in sports, business, and politics. This competition may have prompted Kennedy's tendency as president continually to evoke crisis in his decision making.

According to the second analysis, President Kennedy's assertive self-confidence may have provided an important element of the "Camelot mystique," but carried to a logical conclusion it also led to tense Cold War situations in which on more than one occasion nuclear holocaust became a possible outcome. At the same time, that assertiveness hid a Kennedy weakness for indecisiveness and procrastination until pressed to take a stand. That, coupled with the lack of any essential ideology beyond basic anticommunism and faith in active government, ensured that there was more to the Apollo decision than rational action.

Instead of taking a long view, proponents of the second theory propose, Kennedy engaged in fear-mongering about supposed Soviet strength in space juxtaposed against American weakness, and responded with a lunar-landing decision both spectacular in its reach and outrageous in its cost. Kennedy, because of his competitive nature, was apparently anxious to strap on six-guns and shoot it out with Khrushchev at the OK Corral. This was true despite his recognition that it might not have been the most effective way to deal with the Soviet Union.

A third explanatory approach toward understanding the Apollo decision suggests that Kennedy may have been more oriented toward cooperation with the Soviet Union in space than most people realized. This theory focuses on Kennedy's appeal to Khrushchev in his inaugural address for a cooperative effort in exploring "the stars," and the entreaty in the following state of the union address for a joint development of space research for purposes of weather prediction, communications, and exploration.

Even after Gagarin and the Bay of Pigs, during the month preceding the May 1961 speech announcing Apollo, Kennedy had his brother, Robert F. Kennedy, quietly assess the Soviet leadership's inclinations toward taking a cooperative approach to human space exploration. In addition, NASA Deputy Administrator Hugh L. Dryden undertook a series of talks with Soviet academician Anatoly A. Blagonravov. Kennedy also instructed Wiesner to convene the President's Science Advisory Committee, a panel populated with representatives from outside NASA, to come up with ideas for possible cooperative missions with the Soviets, including an international lunar base. In a memo ironically written the same day as JFK's Apollo speech, Eugene B. Skolnikoff, who was on Wiesner's staff, proposed, "We should offer the Soviets a range of choice as to the degree and scope of cooperation." As Kennedy speechwriter Ted Sorensen remarked, "It is no secret that Kennedy would have preferred to cooperate with the Soviets on space exploration."

Within two weeks of giving his bold May 25 speech, Kennedy met Khrushchev at the Vienna summit and proposed making Apollo a joint mission with the Soviets. The Soviet leader reportedly first said no, then replied, "Why not?" and then changed his mind again, saying that disarmament was a prerequisite for U.S.-USSR cooperation in space. On September 20, 1963, Kennedy made a well-known speech before the United Nations, in which he again proposed a joint human mission to the Moon. He closed by urging, "Let us do the big things together." In public the Soviet Union was noncommittal. *Pravda*, for example, dismissed the 1963 proposal as premature. Some have suggested that Khrushchev viewed the American offer as a ploy to open up Soviet society and compromise Soviet technology. Although these efforts did not produce a cooperative venture—Kennedy was assassinated two months after his U.N. speech and Khrushchev deposed the next year—the fact that Kennedy pursued various forms of space cooperation until his death suggests that he was unsure that a U.S.-only Apollo program was the best course.

Finally, one interpretation of the Apollo decision emphasizes Kennedy as visionary leader—as essentially a space cadet—committed to expanding the human presence throughout the solar system. In this scenario the Apollo decision was merely the first step in an expansive effort to explore and colonize the heavens. Kennedy therefore approved Apollo because he was a visionary who saw space exploration as a noble, worthy goal in its own right. Even without Cold War competition, even without Soviet successes in space, Kennedy would have made his decision to go to the Moon and stuck with it because he considered exploration an objective good. This "romantic" view of space may be attractive—allowing those holding it to cast JFK as a latter-day Columbus or Lewis and Clark—but there is no evidence to support the belief.

Space enthusiasts might like to believe that one of their own occupied the White House and set the nation on a bold spaceflight adventure. Instead, Kennedy maintained a studiously ambivalent record on space exploration before the Gagarin flight of April 1961, neither firmly pro nor con. Journalist Hugh Sidey noted that on assuming the presidency, Kennedy "seemed to know less" and to be "less interested in" space than in virtually any other major policy area.

Finally, recently released tape of a White House meeting on November 21, 1962, between President Kennedy and NASA Administrator Webb demonstrates beyond all dispute the error of the romantic approach. When asked to more aggressively support a broad range of spaceflight activities, Kennedy responded, "I am not that interested in space." The major reason he was expending so much money on Apollo, he said, was because of its importance in the Cold War rivalry with the Soviet Union.

All of this suggests that JFK's Apollo decision was much more complex and involved than most have generally believed. It is, at best, an ambivalent representation of the rational actor approach to decision making in recent American history. In part because of this, the Apollo program left a divided legacy for NASA and the aerospace community. The Apollo decision created for the space agency an expectation that the direction of any major space goal from the president would always bring NASA a broad consensus of support and provide it with the resources and license to dispense as it saw fit. Something NASA officials have been slow to understand is that Apollo had not been conducted under normal political circumstances and would not be repeated.

The Apollo decision was, rather, an anomaly in the national decision-making process. The illusion of the "golden age" of Apollo has been difficult to overcome, and moving beyond the Apollo program to embrace future opportunities has been equally difficult.

JFK and the Limits of Presidential Leadership

In announcing the Moon-landing commitment, President Kennedy had correctly gauged the mood of the nation. His decision was met with strong initial support. No one seemed concerned either about the difficulty or about the expense. Congressional debate was perfunctory and NASA found itself literally pressing to expend the funds committed to it during the early 1960s. Like many political decisions, the decision to carry out Project Apollo was an effort to deal with an unsatisfactory situation, in this case the global perception of Soviet leadership in space and technology. As such, Apollo was a remedial action ministering to a variety of political and emotional needs floating in the ether of world opinion. Apollo addressed these problems effectively, and was worthwhile if measured only in those terms. In announcing Project Apollo, Kennedy put the world on notice that the United States would not take a back seat to its superpower rival. John Logsdon commented: "By entering the race with such a visible and dramatic commitment, the United States effectively undercut Soviet space spectaculars without doing much except announcing its intention to join the contest." It was an effective symbol, just as Kennedy had intended.

It also gave the United States an opportunity to shine. The lunar landing was so far beyond either U.S. or Soviet capabilities in 1961 that the early lead in space activities taken by the Soviets would not predetermine the outcome. Kennedy's commitment gave the United States a reasonable chance of overtaking the USSR in space activities and of recovering a measure of lost status.

Even though Kennedy's political objectives were essentially achieved with the decision to go to the Moon, other aspects of the Apollo commitment require assessment. Those who wanted to see a vigorous space program, a group led by NASA scientists and engineers, obtained their wish with Kennedy's announcement. An opening was present to these partisans in 1961 that had not existed at any time during the Eisenhower administration, and they made the most of it. Into the overall package supporting Apollo, they inserted additional programs that they believed would greatly strengthen the scientific and technological return on the investment to go to the Moon. In addition to seeking international prestige, this group proposed an accelerated and integrated national space effort incorporating both scientific and commercial components.

A unique confluence of political necessity, personal commitment and activism, scientific and technological ability, economic prosperity, and public mood made possible the 1961 decision to carry out an aggressive lunar-landing program. It then fell to NASA and other organizations of the federal government to accomplish the task set out in a few short paragraphs by the president. By the time the goal would be accomplished in 1969, few of the key figures associated with the decision would still be in government leadership positions. Kennedy fell victim to an assassin's bullet in 1963, and Wiesner returned to MIT soon afterward. Lyndon Johnson, of course, succeeded Kennedy as president but left office in January 1969, just a few months before the first landing. Webb resolutely guided NASA through most of the 1960s, but his image was tarnished by, among other things, a 1967 Apollo accident that killed three astronauts; he retired from office under something of a cloud in October 1968. Hugh Dryden and several early supporters of Apollo in Congress died during the 1960s and never saw the program successfully completed.

In some respects, Apollo reflected the peak of what some have called the "imperial presidency." This is the term often given to the aggrandizement of presidential power that came during the administrations of John F. Kennedy, Lyndon B. Johnson, and Richard M. Nixon. The development prompted a number of commentators to criticize the ease with which chief executives overwhelmed other centers of power in the United States. By the time of the Watergate affair, which brought down the Nixon administration in 1974, the expansion and abuse of presidential power relative to the Congress and courts had created a full-blown governmental crisis. Beginning in the mid-1970s, historians and political scientists like Arthur M. Schlesinger, Jr., warned that deference to the president had upset the traditional system of checks and balances.

The Soviets Decide to Race America to the Moon

By May 1961, when the Americans formally announced their intention of landing on the Moon by the end of the 1960s, Soviet aspirations were very real but publicly unstated. When asked about racing the Americans, Soviet officials persistently denied any such objective. Regardless, Korolev, Glushko, Mishin, and many in the Kremlin were intent on besting the Americans in a head-to-head Moon race. Korolev persisted in proposing various schemes to pursue a Moon landing. He recognized the necessity of a large booster designed to hoist seventy-five tons into orbit, and persuaded Khrushchev in 1962 to approve development of the N1 Moon rocket.

In August 1964 Korolev persuaded Khrushchev to back a full-blown lunar-landing program, though responsibility for only part of the effort was assigned to Korolev. Rival design bureaus were brought under Korolev's leadership. The project known as L3 called for the landing of cosmonauts on the Moon before the Americans. The L3 spacecraft would be launched into orbit on the N1 rocket, now with a mandated payload capacity of ninety-five tons. At the same time, Khrushchev gave a Korolev rival, Vladimir Chelomey, a parallel project, known as the LK-1, to send two cosmonauts around the Moon. Korolev and Chelomey thereafter competed for primacy in the space efforts of the Soviet Union and never were able to obtain the resources necessary to be fully successful.

With an alternative space program now operating under Chelomey, another longtime Korolev rival, Valentin Glushko, threw in his lot with Chelomey and built an excellent engine for the LK-1 Moon rocket. With Glushko refusing to support Korolev's Moon effort, Korolev turned to a loyal but untested rocket engine designer

Figure 10. Two N1 Moon rockets appear on the pads at Baikonur in early July 1969. The N1 was designed for the Soviet space program's human lunar missions. In the foreground is booster number 5L with a functional payload for a lunar-orbiting mission. In the background is the IMI ground test mockup of the N1 for rehearsing parallel launch operations. After takeoff the rocket collapsed back onto the pad, destroying the entire pad area in a massive explosion.

named Nikolay Kuznetsov. He tapped Kuznetsov to produce engines for the N1. It proved a huge gamble that did not pay off: Kuznetsov built a fine engine, the NK-33, but it did not have enough power to loft the mighty N1 launcher for a Moon shot. To overcome the lack of lift power, the N1 used thirty NK-33s to power the first stage, and the engineers never were able to get them to work effectively together. Each rocket test was a failure.

Beyond this, infighting took place at every turn. After Khrushchev's overthrow in late 1964, Korolev campaigned to take over Chelomey's circumlunar project. He failed but engineered the cancellation of the LK-1 in October 1965. Korolev accepted the use of a Soyuz spacecraft to send cosmonauts around the Moon powered by a new rocket that Chelomey was developing, the Proton. Korolev promised to undertake a circumlunar mission by late 1967, the fiftieth anniversary of the Bolshevik Revolution. By the end of 1965 the Soviet program was anything but unified; it was planning two distinct components in a circumlunar flyby with cosmonauts and a landing. Both of these were intended to beat the Americans.

Korolev's premature death in January 1966 compounded the confusion. His longtime deputy, Vasily Mishin, took over management but could not contain the rival factions in the Soviet space program. As this took place, the Soviet Union officially denied any Moon-landing program at all.

FIVE

The Game of One-Upmanship

The space race began with a series of space spectaculars, with the Soviet Union enjoying early and sustained success until the mid-1960s. This effort represented a high-water mark of Nikita Khrushchev's premiership, and he exploited it to the fullest. Seemingly, the communist state's rocketeers enabled one stunning success after another until the American Gemini program in 1965, which demonstrated superb American capability. In those first eight years of the space age, it looked as if the Soviet Union did everything right in spaceflight, and the United States appeared at best a weakling without the kind of capabilities that the command economy of the "workers' state" in the Soviet Union had been able to muster. The United States had to mobilize to "catch up" to the apparent might of its Cold War rival. Robert Gilruth, who led the NASA center that handled human spaceflight, and Wernher von Braun, who built the rockets used for Apollo, are central to this story. As surely as the several crises in Berlin—the blockade and airlift, the wall—and the other flashpoints of competition, this game of "one-upmanship" served to fuel the antagonism and steel the resolve of both sides in the Cold War.

Soviet Rocket Development from the R-7 to Soyuz

In both the United States and the Soviet Union a powerful rocket was required to undertake a Moon race. A rocket must be used to overcome the force of gravity holding everything to the surface of this planet. In 1728 British scientist Sir Isaac Newton showed that a cannonball fired from atop a mountain could orbit Earth—that is, if it could travel fast enough and there was no air to slow it down. The bigger the charge used to fire the cannonball, the faster and farther it would travel before Earth's gravity pulled it to the ground. If it achieved enough speed, the downward curve of its path would match the curve of Earth. The cannonball would continue to fall toward the Earth but never actually land—it would be in orbit. Newton calculated orbital speed at 14,500 and 18,000 miles per hour, or between 4 and 5 miles per second, depending on altitude, drag, and other factors.

Harnessing the considerable energy necessary to reach Earth orbit became possible only in the aftermath of technology developments of World War II. However, from orbit it takes far less energy to go someplace else. There is an old saying in the space-exploration community: because of the energy required to escape Earth's gravity, once you reach orbit, you are "halfway to anywhere" you want to go.

While the definition has evolved over time, current international practice defines space as 100 kilometers (62 miles) above Earth. Even 62 miles up, the atmosphere is still dense enough to drag down satellites and spacecraft. They must be boosted to a height approaching 225 miles for normal orbital activities. Space vehicles such as those flown during the space race routinely fly at this altitude.

Building the technology to reach this realm above Earth, and to send spacecraft beyond it, proved challenging. The first to do so were the powerful partners and rivals Sergei Korolev and Valentin Glushko, whose contributions made possible the early successes of the Soviet Union. The large ICBM built by Korolev, the R-7, and powered by Glushko's rocket engines enabled a long list of firsts to be piled up by the Soviet Union.

The R-7, first built as a ballistic missile (SS-6) in the late 1950s but never a practical weapon system because of the lengthy time necessary to prepare it for launch, became famous as the rocket used by Korolev to orbit *Sputniks 1* and *2* as well as the first cosmonauts. It consisted of a core stage surrounded by four strap-on boosters. About two minutes into flight the four strap-ons separated from the core; after the core ran out of fuel, additional upper stages fired.

This became a highly flexible and reliable system for Soviet space launches, and Korolev and Glushko modified it with more advanced upper stages and more efficient engines over time. The version that launched Gagarin in 1961 was known as Block Ye and could orbit about five tons. Another modification with a more powerful upper stage provided the lift for the Voskhod flights of the mid-1960s. Another variant, which became known as Soyuz, began flying in November 1966. Numerous Soyuz variations —Soyuz-L, Soyuz-M, Soyuz-U, and Soyuz-U2—have launched hundreds of payloads into orbit.

American Launchers: Redstone, Atlas, and Titan

While the Korolev/Glushko effort kept to a basic design for its launch vehicles—modifying and upgrading as necessary to provide greater capability—the American approach relied on a succession of designs specific to the mission envisioned. All the early American space race launchers were based on military ballistic missile designs modified for human spaceflight. The Redstone—which launched the first American satellite into orbit and the first two Americans into suborbital flights in 1961—emerged from the rocket team headed by von Braun, leading a cadre of able rocketeers, some of whom had accompanied him from Germany at the end of World War II. Von Braun christened the team's first American missile on April 8, 1952, in recognition of its development at Redstone Arsenal in Huntsville, Alabama, where the U.S. Army concentrated its rocket development activities.

A direct descendant of the German V-2 rocket of World War II, the Redstone's liquid-fueled engine burned alcohol and liquid oxygen and produced about seventy-five thousand pounds of thrust. Nearly seventy feet tall and slightly less than six feet in diameter, it formed the basis of the army's proposal to launch the first human into space even before the establishment of NASA. With the unlikely name of Project Adam, the proposal called for the use of a modified Redstone booster to launch a pilot in a sealed capsule along a steep ballistic, suborbital trajectory. The capsule would reach an altitude of about 150 miles before splashing down by parachute in the Atlantic missile range east of Cape Canaveral, Florida, where von Braun had established launch facilities.

Von Braun and his team developed the Jupiter-C, the first ballistic missile based on the Redstone. The Jupiter-C was an interregional ballistic missile (IRBM) that could destroy transportation crossroads and systems in a war with the Soviet Union from 500 miles away. It could power this small capsule to a speed of thirty-eight hundred miles per hour. For guidance, it used an all-inertial system featuring a gyroscopically stabilized platform, computers, a flight path programmed into the rocket before launch, and the activation of the steering mechanism by signals in flight. For control during powered ascent, the Redstone depended on tail fins with movable rudders and refractory carbon vanes mounted in the rocket exhaust.

The first launch of this rocket took place at Cape Canaveral on

August 20, 1953, and combat-ready troops tested it in battlefield conditions. The Jupiter-C was then placed on active service with U.S. units in Italy and Turkey. The Cold War rivalry between the United States and the USSR had driven the development of rocket technology; it also drove global nuclear fear.

While the rocket was a technical success, deploying it to Italy and Turkey destabilized the Cold War equilibrium with the Soviet Union. President Dwight D. Eisenhower sent the missiles to Europe knowing this: "It would have been better to dump them in the ocean, instead of trying to dump them on our allies." Several national security leaders warned that placement of nuclear weapons so close to the Soviet border was a provocative act that invited attack. Moreover, the technology was such that it took hours to ready the missiles for launch—they had to be deployed at fixed aboveground launch sites, where they could be destroyed by a sniper with a high-powered rifle. The United States deployed them anyway. Khrushchev's gambit in Cuba, where he placed missiles in 1962, was a reaction to these Jupiter missiles. The sides resolved the crisis through back-channel discussions, and each side removed its missiles.

The nuclear fear made real by the Jupiter deployment, and the resulting Cuban missile crisis of October 1962, affected people around the world. The prospect of death in nuclear war prompted many to change their ways of life to deal with the rising threat of nuclear attack. They developed new countermeasures and the protective systems needed to sustain life.

This fear prompted many to build fallout shelters; both Americans and Russians were encouraged to take precautions, and to believe that with enough preparation they could survive a nuclear attack. In addition, the Soviet Union built part of its Moscow Metro system, the Arbatsko-Pokrovskaya Line, exceptionally deep so that it could serve as a collective fallout system for the city. The line housed equipment and supplies to sustain the population during a nuclear attack. Its escalators are some of the longest in the world and move the fastest. The nuclear fear was very real, and while Redstone/Jupiter did not create it, it certainly exacerbated it.

The Redstone version of the rocket, insufficient to send a Mercury capsule into orbit, powered the first two American Mercury flights into space. The missions of Alan Shepard and Gus Grissom in 1961 were suborbital flights of less than thirty minutes. The rocket's virtue for this mission was that it was a well-tested, reliable system that could achieve basic space operations. For orbital Mercury flights a more powerful launcher was needed.

Rather than upgrading Redstone as the Soviets might have done, NASA used an entirely different ballistic missile modified for human spaceflight. A modified Atlas ICBM got the nod. Developed under the leadership of the hard-driving, intense, and flamboyant USAF Brigadier General Bernard A. Schriever, the SM-65 Atlas program officially began in February 1954, under the name Weapon System 107A. The first Atlas rocket was test fired on June 11, 1955, and a later generation rocket became operational in 1959.

When the Atlas was first conceived in the 1950s, many believed it was a high-risk proposition. In order to reduce its weight, Convair Corporation engineers under the direction of Karel J. Bossart, a pre–World War II immigrant from Belgium, designed the booster with a very thin, internally pressurized fuselage instead of massive struts and a thick metal skin. The "steel balloon," as it was sometimes called, employed engineering techniques that ran counter to the conservative engineering approach used by von Braun for the V-2 and the Redstone. Von Braun, according to Bossart, needlessly designed his boosters like "bridges," to withstand any possible shock. For his part, von Braun thought the Atlas too flimsy to hold up during launch. He considered Bossart's approach much too dangerous for human spaceflight, remarking that the astronaut using the "contraption," as he called the Atlas booster, "should be getting a medal just for sitting on top of it before he takes off!" The reservations began to melt away, however, when Bossart's team pressurized one of the boosters and dared one of von Braun's engineers to knock a hole in it with a sledge hammer. The blow left the booster unharmed, but the recoil from the hammer nearly clubbed the engineer.

Once the challenge of mating the Mercury spacecraft to the Atlas was resolved, the stage was set for Glenn's February 1962 flight aboard *Friendship 7* and the three subsequent Mercury flights in 1962 and 1963 powered by the Atlas rocket.

For its second human spaceflight program NASA turned to yet another ICBM program, the Titan rocket, with a capacity to send the heavier and more capable Gemini spacecraft into Earth orbit. In October 1955 the U.S. Air Force contracted with the Glenn L. Martin Company to build the Titan I, the nation's first two-stage ICBM. Designed to be based in underground silos, fifty-four Titan I's were deployed, followed by fifty-four improved Titan II's. The first Titan II ICBMs were activated in 1962, and modified Titan II's were selected to launch NASA's Gemini spacecraft into orbit during the mid-1960s.

"Man-rating" the Titan II ICBM, however, was not straightforward. NASA's Robert Gilruth demanded several modifications to make the Titan II acceptable for human spaceflight:

- Structural modifications for attaching the spacecraft to the launch vehicle;
- Three-axis attitude control, redundant autopilot, and electrical power system;
- Rechargeable space-system batteries in the electrical system;
- Installation of a malfunction detection system (MDS);
- Enhanced propulsion management to solve performance

> reliability and longitudinal oscillation or "POGO" problems and combustion instability problems.

The most serious of these challenges with the Titan II was its longitudinal oscillation, called the POGO effect because it resembled the movements of a child on a pogo stick. Overcoming this problem required engineering imagination and long hours of overtime to stabilize fuel flow and maintain vehicle control. The net result of this effort was an improved Titan II engine system that was "man-rated."

The contrasting approaches of the Soviet Union and the United States to launch technology both were effective. The U.S. approach required more resources, while the Soviet effort built systematically on previous success. The Americans were able to make greater adjustments to their program because of the range of technologies available to them. The Soviets were hamstrung as time passed, because they could do only certain things and not others.

Few understood the limitations of the Soviet capability at the time. In a closed society with successes announced only after the fact, the failures were not usually known to those outside the Soviet leadership. American successes and failures as well became immediately apparent to the world. The one virtue of the Soviets' R-7 rocket also betrayed a weakness of the program overall. The R-7 was built to launch quite large payloads—nuclear warheads—against the United States. Since the USSR lacked the miniaturization capability of the United States, it could not shrink the size and weight of those warheads; accordingly, its rocket had to be larger. That gave the USSR an advantage early on by enabling Korolev to launch the first satellites and humans into space. American launchers initially could not propel as heavy a payload into orbit, but they became more robust over time.

Neither nation at the beginning of the space race had a rocket capable of sending humans to the Moon. Both would have to develop this technology. The American Saturn V, overseen by von Braun and his rocket team, took Americans to the Moon in the latter 1960s and early 1970s. The comparable Soviet N1 rocket was also built but never successfully flew. When Korolev died on January 14, 1966, apparently as a result of complications from a botched hemorrhoid operation, he left a massive hole in the Soviet space program. Competing factions of engineers vied for its control. Korolev's longtime rival, rocket engine designer Valentin Glushko, asserted authority. But no one had the same grasp of technical details of the program's various elements that Korolev had demonstrated, much less the gravitas to unify the program and hold it together. Difficulties dogged the N1 rocket, and after four test launch failures, its development was halted in 1975, more than five years after the Americans reached the Moon.

A Spacecraft Built for Two (or Three): Voskhod and Soyuz

After the success of the early Vostok flights in 1961–1963, Sergei Korolev's design bureau developed the Voskhod spacecraft as a larger, more capable capsule for either two or three cosmonauts. Scaled up from Vostok, the spherical capsule contained the cosmonauts and instruments, and a conical equipment module held engine and propellant. Because of its weight, Korolev used an enhanced R-7 launch vehicle, which later became the basis of the Soyuz booster. Voskhod would later be superseded in 1967 by Soyuz, which with updates and modifications is still used by Russia.

The Soviet Union intended the Voskhod program to explore how the human body reacted to space, but its first two flights satisfied Nikita Khrushchev's thirst for space spectaculars. *Voskhod 1*, October 12–13, 1964, flew three cosmonauts, and *Voskhod 2*, March 18–19, 1965, achieved the first extravehicular activity (EVA), or "spacewalk." After Khrushchev was deposed in October 1964, new USSR leadership shifted away from spaceflights aimed at gaining world prestige, allowed Korolev to cancel the Voskhod program, and put more emphasis on a lunar-landing program.

Voskhod proved, at best, an unimpressive spacecraft, and Korolev chomped at the bit to replace it with something more robust. His lieutenants in OKB-1 worked on a more capable spacecraft named Object 7K or Soyuz (Russian for "union"), initiated in 1962, not long after the U.S. decision to land on the Moon by the end of the decade. Korolev pressed his engineers to build a spacecraft that could send two cosmonauts on a circumlunar flight (Box 3). This proved impractical with the Soyuz capsule, and in 1965 Korolev backtracked to a more modest goal, emphasizing Earth orbital missions of a type and complexity not yet demonstrated by the Americans.

By the fall of 1966 Soyuz was ready for test flights, although Korolev, who had died that January, was not there to see it. The spacecraft that emerged consisted of three modules that weighed about 7.25 tons. The first module contained instruments and service components, as well as electrical and propulsion systems. A habitation module provided accommodation for the crew during its mission, and a small aerodynamic reentry module returned the crew to Earth.

Soyuz made its first flight with a cosmonaut on April 23, 1967; Vladimir Komarov's *Soyuz 1* was intended to rendezvous with a three-cosmonaut *Soyuz 2*, but the mission ran into all manner of problems—not only did *Soyuz 1*'s solar panels fail to extend, causing intermittent electrical outages, but attitude and stability systems also proved disastrously inadequate—and Komarov had to return to Earth. During reentry *Soyuz 1*'s primary and reserve parachutes also failed, causing a crash that killed Komarov on impact, the first Soviet fatality directly related to space exploration and the first inflight death during spaceflight. As his capsule plunged through the

Box 3: Voskhod Specifications

Spacecraft type	
Vostok-3KV	
Vostok-3KD	
Crew capacity	2
Regime	Low Earth
Number Built	5+
Launched	5
Retired	5
First launch	1964
Last launch	1965
Crew size	3 (without space suits)
Service Life	14 days
Overall length	5.0 m (16 feet)
Maximum diameter	2.4 m (8 feet)
Total mass	5,682 kg (6.2 tons)

atmosphere, Komarov could be heard through the radio say, "Heat is rising in the capsule." He also pronounced himself "killed" as he lunged earthward.

After Komarov's death Soviet human space efforts went into hiatus for more than a year and a half. The accident quashed a desire to undertake a space spectacular to commemorate the fiftieth anniversary of the Bolshevik Revolution of October 1917. Not until October 1968 did Soyuz flights resume, with a cosmonaut on *Soyuz 3* trying unsuccessfully to dock with an automated *Soyuz 2*. Only in January 1969 did cosmonauts actually perform a successful mission, with cosmonauts on *Soyuz 4* and *Soyuz 5* docking and spacewalking between the two spacecraft. Thereafter, Soyuz assumed primary operational status for cosmonauts to the present. It has become one of the most successful human spacecraft ever built.

The Vostok, Voskhod, and Soyuz spacecraft provided the technology for the human component of the Soviet space program throughout the 1960s. Many other less-well-known spacecraft were attempted by the Soviets, but none reached the point where cosmonauts flew on them. One of these was called Zond, a robotic spacecraft that some believed could be used to reach the Moon. Zond 5 was launched on September 15, 1968, and took photographs of both Earth and the Moon. Although it flew without cosmonauts aboard, it was capable of carrying them. An attitude control failure put the spacecraft into a ballistic return that would have killed any cosmonauts aboard, but NASA officials wondered whether the Soviets were planning to beat the United States to a circumlunar flight. On November 11, the Soviet Union launched Zond 6, and it also successfully circumnavigated the Moon before returning to Earth. This time the reentry went well, but a gasket failed, and the spacecraft depressurized during descent in Kazakhstan. Even so, NASA leaders became convinced that the next Zond spacecraft might carry cosmonauts on a circumlunar flight. They greenlighted their own Apollo circumlunar mission, *Apollo 8*, which flew in December 1968.

Gemini: The Twins

Meantime, the Americans developed the two-astronaut Gemini capsule, which first flew with astronauts in 1965 and 1966, to learn how to (1) maneuver, rendezvous, and dock with another spacecraft; (2) work outside a spacecraft; and (3) collect physiological data about long-duration spaceflight. To gain experience in these areas before Apollo could be readied for flight, NASA devised Project Gemini. Hatched in the fall of 1961 by engineers at Robert Gilruth's Space Task Group in cooperation with McDonnell Aircraft Corporation technicians, builders of the Mercury spacecraft, Gemini started as a larger Mercury Mark II capsule but soon metamorphosed. It could accommodate two astronauts for extended flights of more than two weeks. The Gemini spacecraft pioneered the use of fuel cells instead of batteries to power the ship, something that would be incorporated into all American human spacecraft thereafter, and it incorporated a series of other modifications to hardware. But problems with the program abounded from the start. The fuel cells that were used to power the spacecraft during flight leaked and had to be redesigned, and the Agena upper stage used for rendezvous and docking required reconfiguration, occasioning costly delays.

One of the objectives for Gemini was to demonstrate a controlled reentry to a preselected landing site. Its designers also toyed with the possibility of using a paraglider being developed at Langley Research Center for "dry" landings instead of a "splashdown" in water and recovery by the navy. This controlled descent and landing was to be accomplished by deploying an inflatable paraglider wing. First NASA built and tested the Paresev, a single-seat, rigid-strut parasail, designed much like a huge hang glider, to test the possibility of a runway landing. The space agency then contracted with North American Aviation (NAA) to undertake a design, development, and test program for a scaled-up spacecraft version of the concept. A full-scale, two-pilot test tow vehicle (TTV) was also built to test the concept and train Gemini astronauts for flight. The TTV tested maneuvering, control, and landing techniques. A helicopter released the TTV, with its wings deployed, over the dry lakebed at Edwards Air Force Base, California, where it safely landed. Scale models of the capsules were released at higher altitudes and faster speeds in order to duplicate reentry conditions.

In operation, the spacecraft would fall through the atmosphere back to Earth when a carefully designed and packed paraglider stowed in the spacecraft deployed, beginning after high-temperature reentry and at subsonic speed, at about 50,000 feet; by 20,000 feet the descending spacecraft would take on the characteristics of a hang glider, and the astronauts would bring the craft to a controlled land-

Figure 11. Godfather to the astronauts Robert R. Gilruth, left, enjoys a moment of levity during the *Gemini 12* mission in 1966. With Gilruth in Houston's Mission Control Center, from right, are astronauts Charles "Pete" Conrad, Jr., John H. Glenn, Jr., and Alan B. Shepard, Jr.

ing on either water or land. Once the wing was deployed, according to a 1963 study of the project, "the pilot directs the vehicle toward a predetermined landing spot by means of manual control in pitch and roll. The pilot executes a flare maneuver at an altitude of some 100 feet above the ground, and the spacecraft lands at a low sink rate." Skids from the spacecraft would serve as landing legs for the crew returning from space.

In all, the research and development program for the paraglider extended from May 1962 through 1965. The latter date was just before the inauguration of human flights of the Gemini spacecraft the following spring. Indeed, as the program proved less successful than originally envisioned, NASA engineers kept pushing back deployment of the paraglider, suggesting that the first few missions could use conventional parachutes. At one point in 1964, NASA wanted to have the first seven Gemini capsules use a traditional parachute recovery system, with the last three missions employing the paraglider. NASA engineers never did get the paraglider to work properly and eventually dropped it from the program in favor of a parachute system like the one used for Mercury.

Even as these trials proceeded, the Gemini program entered its spaceflight phase. Following two unoccupied orbital test flights, the first operational mission took place on March 23, 1965. Mercury astronaut Gus Grissom commanded the mission, with John W. Young, a naval aviator chosen as an astronaut in 1962, accompanying him. The next mission, flown in June 1965, stayed aloft for four days and astronaut Edward H. White II performed the first spacewalk. Eight more missions followed through November 1966 (Table 3). Despite problems great and small encountered on virtually every mission, the program achieved its goals. As a technological learning program Gemini had been a success, with fifty-two experiments performed on the ten missions. The bank of data acquired from Gemini helped to bridge the gap between Mercury and what would be required to complete Apollo within the time constraints directed by the president. Two major objectives had special emphasis during the Gemini program, spacewalking and rendezvous and docking. These proved difficult but the Gemini program allowed NASA astronauts to practice them and ultimately master the techniques. In both spacewalking and rendezvous and docking, the Soviet program also staged major firsts in the game of one-upmanship with the Americans.

Table 3

Gemini Missions, 1965–1966

Mission	Launch Date	Crew	Flight Time (days:hrs:mins)	Highlights
Gemini 3	Mar. 23, 1965	Virgil I. Grissom John W. Young	0:4:53	First U.S. 2-person flight; first manual maneuvers in orbit
Gemini 4	Jun. 3, 1965	James A. McDivitt Edward H. White	4:1:56	21-min extravehicular activity (White)
Gemini 5	Aug. 21, 1965	L. Gordon Cooper Charles Conrad, Jr.	7:22:55	Longest-duration human flight to date
Gemini 7	Dec. 4, 1965	Frank Borman James A. Lovell, Jr.	13:18:35	Longest human flight to date
Gemini 6-A	Dec. 15, 1965	Walter M. Schirra Thomas P. Stafford	1:1:51	Rendezvous within 12 inches of *Gemini 7*
Gemini 8	Mar. 16, 1966	Neil A. Armstrong David R. Scott	0:10:41	First docking of 2 orbiting spacecraft (*Gemini 8* with an Agena target vehicle)
Gemini 9-A	Jun. 3, 1966	Thomas P. Stafford Eugene A. Cernan	3:0:21	Extravehicular activity; rendezvous
Gemini 10	Jul. 18, 1966	John W. Young Michael Collins	2:22:47	First dual rendezvous (*Gemini 10* with *Agena 10* and *Agena 8*)
Gemini 11	Sep. 12, 1966	Charles Conrad, Jr. Richard F. Gordon, Jr.	2:23:17	First initial-orbit docking; first tethered flight; highest Earth-orbit altitude (850 miles)
Gemini 12	Nov. 11, 1966	James A. Lovell, Jr. Buzz Aldrin	3:22:35	Longest extravehicular activity to date (Aldrin, 5 hrs, 37 mins)

Source: NASA, *Aeronautics and Space Report of the President, 1974 Activities* (Washington, D.C.: NASA, 1975), appendix C, 137–139.

Spacewalking

Both the Americans and the Soviets realized early that true space exploration required astronauts and cosmonauts to be able to leave the spacecraft. In 1965 both national space programs undertook the first spacewalks. With the American Gemini program under way and a slate of announced objectives in place—one of which was spacewalking—Soviet cosmonaut Alexei Leonov beat the United States to the first spacewalk during the *Voskhod 2* mission on March 18, 1965. Leonov commented: "What struck me most was the silence. It was a great silence, unlike any I have encountered on Earth, so vast and deep that I began to hear my own body: my heart beating, my blood vessels pulsing, even the rustle of my muscles moving over each other seemed audible. There were more stars in the sky than I had expected. The sky was deep black, yet at the same time bright with sunlight."

He pushed away from the vehicle and drifted out 17.5 feet before returning to the spacecraft. A tense few moments ensued when Leonov found that his space suit had ballooned when pressurized and was too rigid to reenter the airlock. He solved the problem by bleeding air from his suit to reduce its size. As Leonov wrote about this experience:

> During my training for this mission, I did a drawing showing how I imagined myself walking in space high over the planet Earth in the outer cosmos. The dream came true, and space walking became a reality with my EVA on Voskhod 2 in March 1965. During the space walk, I was exposed to the vacuum of space for some 20 minutes, considerably longer than expected, due to problems re-entering the spaceship. The pressure difference between air in my space suit and the vacuum of the cosmos expanded my space suit and made it rigid, and I had to force some of the air out of the suit in order to close the lock's outer hatch.

The Soviet spacewalk was a blatant attempt to steal the march on the American Gemini effort. There was little accomplished other than the distinction of being first, however, and the mission did not lead to any sustained efforts thereafter. Indeed, the Soviets did not undertake another spacewalk until Yevgeny Khrunov and Aleksei Yeliseyev performed the first two-person spacewalk on January 16, 1969, during the *Soyuz 4* and *5* missions.

The first American spacewalk came on June 3, 1965, when Astronaut Ed White floated out of the *Gemini 4* capsule. That was a simple affair, floating in space, but NASA soon learned that performing useful functions while weightless was more challenging. The second Gemini spacewalk nearly proved deadly. On June 5, 1966, astronaut Eugene Cernan departed the Gemini 9-A spacecraft in Earth orbit to conduct a pivotal extravehicular activity. It was only the second time an American astronaut had ever ventured outside of a capsule to expose the body to the extreme environment of space. Cernan quickly learned that anything he did in microgravity took more energy than anticipated, and his body overheated. This overtaxed his space suit's environmental system. His helmet visor fogged over, making it impossible to see, sweat poured into his eyes, and his heart raced to more than three times its normal rate. He lost nearly ten pounds from dehydration. Finally, after more than two hours, and one and a half orbits of Earth, a drained Gene Cernan made it back inside his Gemini spacecraft after failing to complete most of the objectives of his EVA.

At the same time, NASA learned valuable lessons both about the fragility of the human body in the extreme environment of space and the care to be taken to accomplish useful work while in zero gravity. Its engineers redesigned space suits to provide more robust life support. Additionally, led by Buzz Aldrin, the astronauts developed procedures to more effectively conduct useful work while weightless. Aldrin's work on spacewalking made it possible to depart the spacecraft and do something useful in the vacuum. Systematically and laboriously, Aldrin worked to develop procedures necessary to conduct spacewalks. He went into a large swimming pool to simulate the experience of zero gravity, and this step of the testing taught engineers how to modify the space suit to keep it from hindering movement by ballooning up and to keep it from either superheating or chilling the astronaut inside. Aldrin eventually confirmed the soundness of the new suit and the procedures developed for EVA during *Gemini 12* in November 1966, spending more than five hours outside the spacecraft and performing several functions necessary for the Moon landings.

Likewise, the rendezvous and docking of two spacecraft proved more difficult than envisioned. Holding a Ph.D. in astronautics from the Massachusetts Institute of Technology, Aldrin again played a significant role in developing the necessary procedures. His dissertation had offered the theoretical basis for orbital rendezvous, and it provided the basis for the maneuvering and docking of two spacecraft in Earth orbit. Demonstration of this task proved successful only after many failures. For example, on March 16, 1966, Neil A. Armstrong flew his first space mission as command pilot of *Gemini 8* with David Scott. During that mission, Armstrong piloted the spacecraft to a successful docking with an Agena target spacecraft already in orbit. Although the docking went smoothly and the two spacecraft orbited together, they began to pitch and roll wildly. Armstrong undocked the Gemini and used retrorockets to regain control of his craft, but the astronauts had to make an emergency landing in the Pacific Ocean. After subsequent trials, however, by the end of the Gemini program rendezvous and docking was routine activity that astronauts performed as a part of the normal mission duties. Soviet cosmonauts did not have similar success until much later. Indeed, all their spacecraft were remotely piloted for docking maneuvers until the era of the International Space Station in the twenty-first century.

World Opinion Shifts

Throughout the first years of the space age, public opinion in the United States gave weight to the demonstrations of success made by the USSR. The Soviets' list of accomplishments was long, and most Americans thought the game of one-upmanship was being won by the Soviets. This was borne out in responses to public opinion polls on the question (Graph 1). For most of the time during the late 1950s and at the start of the 1960s, the Soviets were viewed as in the lead. This perception began to shift in the middle part of the decade as the Gemini program began to come to fruition. It never relented thereafter. By the completion of the Moon race the United States was viewed globally as the leader in science and technology to which all other nations wanted to be attached.

American concerns about this perceived lead by the Soviet Union in space sparked the decision to send American astronauts to the Moon during the 1960s, of course, but it manifested itself in a fundamental manner in terms of swaying support toward the American geopolitical coalition. The competition in space revolved, first, around ensuring that allies in the Cold War struggle with the Soviet Union remained in the American camp. Second, lassoing newly emerging nations—especially those recently achieving their independence from European colonial rule—into the American orbit. The fear that there would be more nations like Cuba, whose 1959 leftist revolution led by Fidel Castro had established a Soviet beachhead in the Western Hemisphere and led to years of covert operations against him, permeated American thinking. Preventing that eventuality by impressing these nations with American technologi-

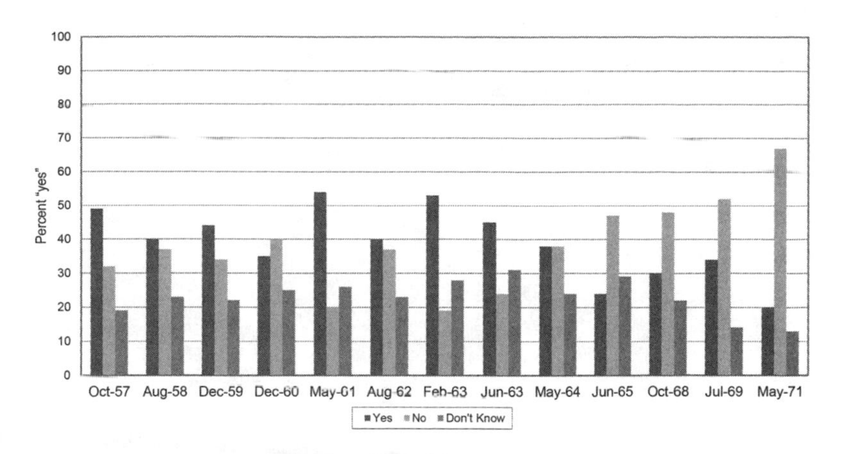

Graph 1. Is Russia Ahead of the United States in Space?
Source: Gallup Polls; wording of questions differed slightly.

cal verisimilitude became a major part of what American strategy for the space race engendered.

Effective leaders of emerging nations such as Indira Gandhi in India played the United States and the USSR against each other. In an effort to obtain the best results for her advancing nation, especially in technological and economic development, Gandhi made deals with both sides. Enthralled by early success in space exploration by the Soviet Union and the United States, Indian scientist Vikram Sarabhai sought relationships and cooperative space efforts with both powers. He engineered the creation of the Indian National Committee for Space Research (INCOSPAR) in 1962 and its successor the Indian Space Research Organization (ISRO) in 1969. India worked with the Soviet Union to develop and launch its own satellite, *Aryabhata*, which orbited on April 19, 1975, to conduct experiments in X-ray astronomy and solar physics, and to measure ionized gas in the upper atmosphere. At the same time India's scientists

Table 4

First Ten Nations in Space after USA/USSR, 1962–1974

Country	Satellite	Operator	Launcher	Launch Site	Launch Date
U.K. (joint USA)	*Ariel 1*	Royal Aircraft Establishment	Thor/Delta	Cape Canaveral, USA	April 26, 1962
Canada (joint USA)	*Alouette 1*	Defence Research and Development Canada	Thor/Agena-B	Vandenberg AFB, USA	September 29, 1962
Italy (joint USA)	*San Marco 1*	Commissione per le Ricerche Spaziali (CRS)	Scout X-4	Wallops Island, USA	December 15, 1964
France	*Astérix*	Centre national d'études spatiales (CNES)	Diamant A	Hammaguir, Algeria	November 26, 1965
Australia	*WRESAT*	Weapons Research Establishment (WRE)	Sparta	Woomera, Australia	November 29, 1967

Europe	*ESRO 2B*	European Space Research Organisation (ESRO)	Scout B	Vandenberg AFB, USA	May 17, 1968
Federal Republic of Germany (joint USA)	*Azur*		Scout B	Vandenberg AFB, USA	November 8, 1969
Japan	*Ōsumi*	Institute of Space and Aeronautical Science (ISAS)	Lambda-4S	Kagoshima, Japan	February 11, 1970
People's Republic of China	*Dong Fang Hong 1*		Chang Zheng 1	Jiuquan, China	April 24, 1970
The Netherlands (joint USA)	*Astronomical Netherlands Satellite (ANS)*		Scout	Vandenberg AFB, USA	August 29, 1974

Sources: NASA, *Aeronautics and Space Report of the President, 1974 Activities* (Washington, D.C.: NASA, 1975), appendix A-2, 125–126; Roger D. Launius, *The Smithsonian History of Space Exploration* (Washington, D.C.: Smithsonian Books, 2018), 170–171.

and engineers built strong ties to NASA for cooperative projects and were involved as key investigators on several scientific probes launched by NASA in the 1960s and 1970s.

Other nations came into the American sphere of space exploration in the latter half of the 1960s, as perceptions began to shift in favor of the demonstrated success of the Gemini program and the achievement of the Moon-landing effort. A case in point was Pakistan, which since 1961 had been eager to enter the "space club," a loose confederation of nations engaging in spacefaring activities, as a means of demonstrating that it deserved a seat at the table of world governance. Impressed by both American and Soviet successes, Pakistan eventually entered agreements with the United States to cooperate on scientific efforts. These examples were repeated many times around the globe, and from the time of the first cooperative ventures with American allies in the 1960s, a multination consortium of efforts blossomed during the post–space race era of the decades that followed (Table 4).

Could the Moon Race Have Become a Cooperative Program?

John F. Kennedy's decision to send Americans to the Moon by the end of the decade, announced in May 1961, was a visible challenge to the Soviet Union to demonstrate technological supremacy in space exploration. The winner would gain prestige among the nonaligned nations of the world, perhaps swaying them into the winner's Cold War coalition. The competition was real, it was impressive, and it opened the treasury to achieve that greatest of all endeavors, setting foot on another body in the solar system. And the Soviet Union took Kennedy's bait, beginning its own aggressive effort in 1964 to reach the Moon before the Americans.

But what if the United States and the USSR had undertaken the Moon-landing program cooperatively rather than as a competition?

It is more than an academic question. As we have seen, there were genuine efforts to make it a joint program, from JFK's appeal to Khrushchev in his inaugural address almost to the end of Kennedy's truncated term in office. Of course, no cooperative venture materialized. What might have been had Kennedy lived, had Khrushchev been able to hold on to power? After they were gone, the game of one-upmanship continued, but as the sixties progressed, the Soviet Union registered fewer successes. In part that was because the Americans had developed and were systematically working to achieve a Moon landing, in part it was because the towering presence of Sergei Korolev had passed from the scene and no one else was able to hold together the competing factions inside the Soviet technical bureaucracy. The result was a race to the finish, but the only nation making a creditable effort was America. NASA had a national mandate, sufficient resources, and clear direction; it completed an impressive Moon-landing program during the 1969–1972 era.

SIX

Creating the Moon-Landing Capability

The race to the Moon that the United States and the Soviet Union undertook led to sophisticated expressions of technological capability on both sides. Charting these developments consumes this chapter, and both national efforts certainly registered positive results in the face of unique challenges and tragic accidents. Putting a human on the Moon in 1969 was a feat of astounding technological virtuosity. Landing there six times, as the Americans did, was overwhelmingly significant. Project Apollo succeeded because it was a triumph of management in meeting enormously difficult systems engineering and technological integration requirements.

Gearing Up for Project Apollo

The first challenge NASA leaders faced in meeting the presidential mandate in 1961 of a Moon landing was to secure resources. While Congress enthusiastically appropriated funding for Apollo immediately after the president's announcement, NASA Administrator James E. Webb was rightly concerned that the momentary sense of crisis would subside and that the political consensus present for Apollo in 1961 would abate. He tried, albeit without much success, to lock the presidency and the Congress into a long-term pledge to support the program. While the politicians had made an intellectual commitment, NASA's leadership was concerned that they might later renege on the economic part of the bargain.

Initial NASA estimates of the costs of Project Apollo were about $15 billion through the end of the 1960s, a figure of nearly $200 billion in 2018 dollars. Webb quickly stretched those initial estimates for Apollo as far as possible, sometimes evening doubling them, with the intent that even if NASA did not receive its full budget requests, as it did not during the latter half of the decade, it would still be able to complete Apollo. At one point in 1963, for instance, Webb highballed estimates for Apollo funding through 1970 at more than $35 billion. This was just to give breathing space for the program if necessary. A tally of Apollo costs presented by NASA to Congress in 1973 stood at $25.4 billion in current dollars. That remained NASA's official characterization of Apollo costs until a 1979 analysis, presented in Graph 2, revised the number downward to $21,347,641, with the single highest expenditure in any year coming in 1969, the year of the first two Moon landings. As it turned out, by consistently overestimating the costs, Webb was able to sustain the momentum of Apollo through the decade, largely because of his rapport with key members of Congress and with Lyndon B. Johnson, who became president in November 1963.

Project Apollo, backed by sufficient funding, was the tangible result of an early national commitment in response to a perceived threat to the United States by the Soviet Union. NASA leaders recognized that while the size of the task was enormous, it was still technologically and financially within their grasp, but they had to move forward quickly. Accordingly, the space agency's annual budget increased from $500 million in 1960 to a high point of $5.2 billion

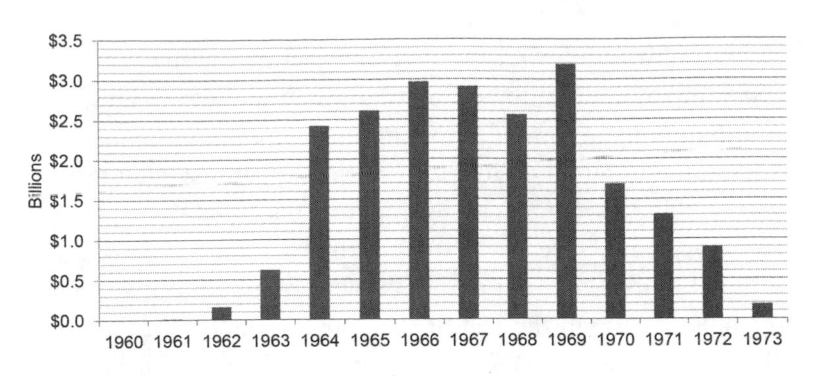

Graph 2. Apollo Costs by Year
In billions of contemporary dollars. The official estimate provided to Congress in 1973 was $25.4 billion.
Sources: Chariots for Apollo: A History of Manned Lunar Spacecraft (NSA SP-4205, 1979), appendix H, 409–411; House Subcommittee on Manned Space Flight of the Committee on Science and Astronautics, 1974 NASA Authorization, Hearings on H.R. 4567, 93/2, part 2, p. 1271.

in 1965 (Table 5). The NASA funding level represented 5.3 percent of the federal budget in 1965. A comparable percentage of the $4.09 trillion federal budget in 2018 would have equaled more than $216 billion for NASA, whereas the agency's actual budget then stood at less than $20 billion. The money spent on Apollo was a significant amount to be sure, but GDP, a measure of the total output of the American economy, for 1965 was nearly $4.3 trillion. NASA's share of that amount was 0.07 percent—7 one hundredths of a percent. The nation could afford to go to the Moon even if it was a significant investment.

Out of the budgets appropriated for NASA each year approximately 50 percent went directly for human spaceflight, and the vast majority of that went directly toward Apollo. Between 1959 and 1973 NASA spent $23.6 billion on human spaceflight, exclusive of infrastructure and support, of which nearly $20 billion was for Apollo. In addition, Webb sought to expand the definition of Proj-

Table 5
Overall NASA Budget in Billions of Dollars
(Uncorrected for Inflation)

Year	Amount	Percentage Change
FY 60	$0.500	
FY 61	$0.964	93%
FY 62	$1.825	89%
FY 63	$3.674	101%
FY 64	$5.100	39%
FY 65	$5.250	2.9%
FY 66	$5.175	–1.4%
FY 67	$4.966	–4.0%
FY 68	$4.587	–7.6%
FY 69	$3.991	–13%
FY 70	$3.746	–6.1%
FY 71	$3.311	–12%

Sources: NASA, *Aeronautics and Space Report of the President, 1974 Activities* (Washington, D.C.: NASA, 1975), appendix E-1 and E-2, 140–141.

ect Apollo beyond just the mission of landing humans on the Moon. As a result, even those projects not officially funded under the Apollo line item, such as the Ranger, Lunar Orbiter, and Surveyor satellite probes, could be justified as supporting the mission.

For seven years after Kennedy's Apollo decision, through October 1968, James Webb politicked, coaxed, cajoled, and maneuvered for NASA in Washington. A longtime Washington insider—the former director of the Bureau of the Budget and undersecretary of state during the Truman administration—he was a master at bureaucratic politics, understanding that it was essentially a system of give and take. For instance, whether the native North Carolinian genuinely believed in the Johnson administration's civil rights bill that went before Congress in 1964, as a personal favor to the president he lobbied for its passage on Capitol Hill. This secured for him Johnson's gratitude, which he then used to secure the administration's backing of NASA's initiatives. In addition, Webb wielded the money appropriated for Apollo to build up a constituency for NASA that was both powerful and vocal. This type of gritty pragmatism also characterized Webb's dealings with other government officials and members of Congress throughout his tenure as administrator. When give and take did not work, as was the case on occasion with some members of Congress, Webb used the presidential directive as a hammer to get his way. Usually this proved successful. After Kennedy's assassination in 1963, moreover, he sometimes appealed for continued political support for Apollo because it represented a fitting tribute to the fallen leader. In the end, through a variety of methods Administrator Webb built a seamless network of political liaisons that brought continued support for and resources to accomplish the Apollo Moon landing on the schedule Kennedy had announced.

Funding was not the only critical component for Project Apollo. To realize the goal of Apollo under the strict time constraints mandated by the president, personnel had to be mobilized. This took two forms. First, by 1966 the agency's civil service rolls had grown to 36,000 people from the barely 8,000 employed at NASA in 1958. Additionally, NASA's leaders made an early decision that they would have to rely upon outside researchers and technicians to complete Apollo, and contractor employees working on the program increased by a factor of 10, from 36,500 in 1960 to 376,700 in 1965 (Table 6). Private industry, research institutions, and universities provided most of the personnel working on the space effort.

To incorporate the great amount of work undertaken for the project into the formal bureaucracy never seemed a particularly savvy idea, and as a result during the 1960s somewhere between 80 and 90 percent of NASA's overall budget went for contracts to purchase

Table 6
NASA Personnel Mobilization

Year	In-House NASA	Contractor Employees	Ratio
1958 (9/30)	8,040		
1960	10,200	36,500	1:3.6
1961	17,500	57,000	1:3.3
1962	23,700	115,500	1:4.9
1963	29,900	218,400	1:7.3
1964	32,500	347,100	1:10.7
1965	34,300	376,700	1:11
1966	36,000	360,000	1:10
1967	36,200	272,900	1:7.5
1968	35,000	211,200	1:6
1969	34,000	200,000	1:5.9
1970	32,500	170,000	1:5.2
1971	30,500	130,000	1:4.3

Sources: NASA Historical Data Book, 1958–1968, vol. 1, *NASA Resources* (Washington, D.C.: NASA SP-4012, 1976); *NASA Pocket Statistics* (Washington, D.C., January 1971); Personnel Analysis and Evaluation Office, NASA Headquarters, Washington, D.C., May 1986; Sylvia D. Fries, *NASA Engineers and the Age of Apollo* (Washington, D.C.: NASA SP-4104, 1992), Appendix B.

goods and services from business firms and universities. This reliance on the private sector and universities for the bulk of the effort predated Apollo, originating early in NASA's history under T. Keith Glennan, in part because of the Eisenhower administration's mistrust of large government establishments. Although neither Glennan's successor nor Kennedy shared that mistrust, they found that it was both good politics and the best way of getting Apollo done on the president's schedule. It was also very nearly the only way to harness talent and institutional resources already in existence in the emerging aerospace industry and the country's leading research universities.

In addition to these other resources, NASA moved quickly during the early 1960s to expand its physical capacity so that it could accomplish Apollo. In 1960 the space agency consisted of a small headquarters in Washington, its three inherited NACA research centers, and the Jet Propulsion Laboratory, the Goddard Space Flight Center, and the Marshall Space Flight Center. With the advent of Apollo, these installations grew rapidly. In addition, NASA added three facilities specifically to meet the demands of the lunar-landing program. In 1962 it created the Manned Spacecraft Center (renamed the Lyndon B. Johnson Space Center in 1973) near Houston to design the Apollo spacecraft and the launch platform for the lunar lander. This center also became the home of NASA's astronauts and the site of mission control. NASA then greatly expanded the Launch Operations Center at Cape Canaveral on Florida's eastern seacoast. Renamed the John F. Kennedy Space Center on November 29, 1963, this installation's massive and expensive Launch Complex 39 was the site of all Saturn V firings. The spaceport's Vertical Assembly Building was a huge and expensive forty-six-story structure where the Saturn/Apollo rockets were assembled. Finally, to support the development of the Saturn launch vehicle, in October 1961 NASA created on a Deep South bayou the Mississippi Test Facility, renamed the John C. Stennis Space Center in 1988. The cost of this expansion was great, more than $2.2 billion over the decade, with 90 percent of it expended before 1966.

The Program Management Concept

The mobilization of resources was not the only challenge facing those charged with meeting President Kennedy's goal. NASA had to meld disparate institutional cultures and approaches into an inclusive organization moving along a single unified path. Each NASA installation, university, contractor, and research facility had its own perspective on how to go about the task of accomplishing Apollo. To bring a semblance of order to the program, NASA expanded the "program management" concept borrowed by Glennan in the late 1950s from the military-industrial complex, bringing in military managers to oversee Apollo. The central figure in this process was U.S. Air Force Major General Samuel C. Phillips, the architect of the Minuteman ICBM program before coming to NASA in 1962. Answering directly to the Office of Manned Space Flight at NASA headquarters, which in turn reported to the NASA administrator, Phillips created an omnipotent program office with centralized authority over design, engineering, procurement, testing, construction, manufacturing, spare parts, logistics, training, and operations.

One of the fundamental tenets of the program management concept was that three critical factors—cost, schedule, and reliability—were interrelated and had to be managed as a group. Many also recognized the consistent interrelation of these factors: if program managers held cost to a specific level, then one of the other two factors, or both of them to a lesser degree, would be adversely affected. The Apollo schedule, dictated by the president, was firm. Since humans were involved in the flights, and since the president had directed that the lunar landing be conducted safely, the program managers placed a heavy emphasis on reliability. Accordingly, Apollo used redundant systems extensively so that failures would be both predictable and minor in consequences. The significance of both of these factors forced the third factor, cost, much higher than might have been the case with a more leisurely lunar program such as had been conceptualized in the latter 1950s. As it was, this was the price paid for success under the Kennedy mandate, and program managers made conscious decisions based on these factors.

The program management concept was recognized as a critical component of Project Apollo's success in November 1968, when

Science magazine, the publication of the American Association for the Advancement of Science, observed:

> In terms of numbers of dollars or of men, NASA has not been our largest national undertaking, but in terms of complexity, rate of growth, and technological sophistication it has been unique. . . . It may turn out that [the space program's] most valuable spin-off of all will be human rather than technological: better knowledge of how to plan, coordinate, and monitor the multitudinous and varied activities of the organizations required to accomplish great social undertakings.

Understanding the management of complex structures for the successful completion of a multifarious task was an important outgrowth of the Apollo effort.

This management concept under Phillips orchestrated more than 500 contractors working on both large and small aspects of Apollo. For example, recipients of the prime contracts awarded to industry for the principal components of just the Saturn V included the Boeing Company for the S-IC, first stage; North American Aviation for the S-II, second stage; the Douglas Aircraft Corporation for the S-IVB, third stage; the Rocketdyne Division of North American Aviation for J-2 and F-1 engines; and International Business Machines (IBM) for the Saturn instruments. These prime contractors, with more than 250 subcontractors, provided millions of parts and components for use in the Saturn launch vehicle, all meeting exacting specifications for performance and reliability. The total cost expended on development of the Saturn launch vehicle was $9.3 billion. So huge was the overall Apollo endeavor that NASA's individual contracts for all things purchased great and small rose from roughly 44,000 in 1960 to almost 300,000 by 1965.

Getting all the personnel elements, including civil service, industry, and university personnel, to work together challenged the program managers. Various communities within NASA differed over priorities and competed for resources. The two most identifiable groups were the engineers and the scientists. As ideal types, engineers usually worked in teams to build hardware that could by the end of the decade carry out the missions necessary to a successful Moon landing. Their primary goal involved building vehicles that would function reliably within the fiscal resources allocated to Apollo. Again as ideal types, space scientists engaged in pure research and were more concerned with designing experiments that would expand scientific knowledge about the Moon. They also tended to be individualists, unaccustomed to regimentation and unwilling to concede gladly the direction of projects to outside entities. The two groups contended with each other over a great variety of issues associated with Apollo. For instance, the scientists disliked having to configure payloads so that they could meet time, money, or launch vehicle constraints. The engineers, likewise, resented changes to scientific packages added after project definition because these threw their hardware efforts out of kilter. Both had valid concerns, but they had to maintain an uneasy cooperation to accomplish Project Apollo.

Furthermore, neither the scientific nor the engineering community within NASA was monolithic, and differences within each thrived. Add to these groups representatives from industry, universities, and research facilities, and competition on all levels to further individual scientific and technical areas was inevitable. The NASA leadership generally viewed this pluralism as a positive force within the space program, for it ensured that all sides aired their views and honed their positions to a fine edge. Competition, most people concluded, made for a more precise and viable space-exploration effort. There were winners and losers in this strife, however, and sometimes ill will festered for years. Moreover, if the debates between scientists and engineers, even between scientists or between engineers with differing priorities, became too great and spilled into areas where they were misunderstood, it could be devastating to the conduct of the lunar program. Most important, disagreements between those inside NASA could never be fought out in public settings such as congressional hearings without raising concerns among the rank-and-file public that the presumed experts at NASA might not really know what they were doing. Disagreements, NASA leaders emphasized, needed to be aired and resolved inside the agency and not on the nightly news or in the newspapers. The head of the Apollo program worked hard to keep these factors balanced and to promote order so that NASA could accomplish the presidential directive.

Another important management issue arose from the agency's inherited culture of in-house research. Because of the magnitude of Project Apollo, and its compressed schedule, most of the nitty-gritty work had to be done outside NASA by means of contracts. As a result, with a few important exceptions, NASA scientists and engineers did not build flight hardware, or even operate missions. Rather, they planned the program, prepared guidelines for execution, evaluated competing contract bids, and oversaw work accomplished elsewhere. This grated on those NASA personnel oriented toward research, and prompted disagreements over how to carry out the lunar-landing goal. Nor were the complaints just that NASA personnel wanted to be "dirty-handed" engineers; NASA scientists and engineers needed to have enough in-house expertise to ensure program accomplishment. If in-house personnel lacked professional competence on a par with that of the individuals actually doing the work, how could NASA oversee contractors actually creating the hardware and performing the experiments necessary to meet the rigors of the mission?

One anecdote illustrates this point. The Saturn second stage was built by North American Aviation at its plant at Seal Beach, California, shipped to NASA's Marshall Space Flight Center in Huntsville, Alabama, and there tested to ensure that it met contract specifications. Problems arose on this piece of the Saturn effort, and Wernher von Braun began intensive investigations. Essentially his engineers completely disassembled and examined every part of every stage delivered by North American to ensure that no defects remained. This was an enormously expensive and time-consuming process, grinding the stage's production schedule almost to a standstill and jeopardizing the presidential timetable.

When this happened, Webb told von Braun to desist, adding, "We've got to trust American industry." The issue came to a showdown at a meeting where the Marshall rocket team was asked to explain its extreme measures. One of the engineers produced a rag and told Webb, "This is what we find in this stuff." The contractors, the Marshall engineers believed, required extensive oversight to ensure that they produced the highest-quality work. A compromise emerged that was called the 10 percent rule: 10 percent of all funding for NASA was to be spent to ensure in-house expertise and in the process to check contractor reliability.

How Do We Go to the Moon?

One of the critical early management decisions made by NASA was the method of going to the Moon. No controversy in Project Apollo more significantly illustrated the tenor of competing constituencies in NASA than this one. There were three basic approaches that were advanced to accomplish the lunar mission:

1. *Direct ascent* called for the construction of a huge booster that launched a spacecraft, sent it on a course directly to the Moon, landed a large vehicle, and sent some part of it back to Earth. The Nova booster project, which was to have been capable of generating up to 40 million pounds

of thrust, would have been able to accomplish this feat. Even if other factors had not impaired the possibility of direct ascent, the huge cost and technological sophistication of the Nova rocket quickly ruled out the option and resulted in cancellation of the project early in the 1960s, despite the conceptual simplicity of the direct ascent method. The method had few advocates when serious planning for Apollo began.

2. *Earth-orbit rendezvous* was the logical first alternative to the direct ascent approach. It called for the launching of various modules required for the Moon trip into an orbit above Earth, where they would rendezvous, be assembled into a single system, refueled, and sent to the Moon. This could be accomplished using the Saturn launch vehicle already under development by NASA and capable of generating 7.5 million pounds of thrust. A logical byproduct of this approach would be the establishment of a space station in Earth orbit to serve as the lunar mission's rendezvous, assembly, and refueling point. In part because of this prospect, a space station emerged as part of the long-term planning of NASA as a jumping-off place for the exploration of space. This method of reaching the Moon, however, was also fraught with challenges, notably finding methods of maneuvering and rendezvousing in space, assembling components in a weightless environment, and safely refueling spacecraft.
3. *Lunar-orbit rendezvous* entailed sending the entire lunar spacecraft up in one launch. It would head to the Moon, enter into orbit, and dispatch a small lander to the lunar surface. It was the simplest of the three methods, in terms of both development and operational costs, but it was risky. Since rendezvous was taking place in lunar, instead

> of Earth, orbit, there was no room for error; any slipup, and the crew could not get home. Moreover, some of the trickiest course corrections and maneuvers had to be done after the spacecraft had been committed to a circumlunar flight. The Earth-orbit rendezvous approach kept all the options for the mission open longer than the lunar-orbit rendezvous mode.

Inside NASA, advocates of the various approaches contended over the method of flying to the Moon while the all-important clock that Kennedy had started continued to tick. It was critical that a decision not be delayed, because the mode of flight in part dictated the spacecraft to be developed. While NASA engineers could proceed with building a launch vehicle, the Saturn, and define the basic components of the spacecraft—a habitable crew compartment and a jettisonable service module containing propulsion, electronics, and other expendable systems—they could not proceed much beyond rudimentary conceptions without a mode decision. The NASA Rendezvous Panel at Langley Research Center, headed by John C. Houbolt, pressed hard for the lunar-orbit rendezvous as the most expeditious means of accomplishing the mission. Using sophisticated technical and economic arguments, over a period of months in 1961 and 1962 Houbolt's group advocated and persuaded the rest of NASA's leadership that lunar-orbit rendezvous was not the risky proposition that it had earlier seemed.

The last to give in was von Braun and his associates at the Marshall Center. This group favored the Earth-orbit rendezvous because the direct ascent approach was technologically unfeasible before the end of the 1960s, because it provided a logical rationale for a space station, and because it ensured an extension of the Marshall workload (something that was always important to center directors competing inside the agency for personnel and other resources). At an all-day meeting on June 7, 1962, at Marshall, NASA leaders met to hash out these differences, with the debate getting heated at times. After more than six hours of discussion, von Braun finally gave in to the lunar-orbit rendezvous mode, saying that its advocates had demonstrated adequately its feasibility and that any further contention would jeopardize the president's timetable.

With internal dissention quieted, NASA moved to announce the Moon-landing mode to the public in the summer of 1962. As NASA prepared to do so, however, Kennedy's science adviser, Jerome B. Wiesner, raised objections because the lunar-orbit rendezvous approach brought the potential risk of losing a crew that could not return to Earth. As a result of this opposition, Webb backpedaled and stated that the decision was tentative and that NASA would sponsor further studies. The issue reached a climax at Marshall in September 1962 when President Kennedy, Wiesner, Webb, and several other Washington figures visited von Braun. As the entourage viewed a mockup of a Saturn V first-stage booster during a photo opportunity for the media, Kennedy nonchalantly mentioned to von Braun, "I understand you and Jerry disagree about the right way to go to the moon."

Von Braun acknowledged this disagreement, but when Wiesner began to explain his concern Webb, who had been quiet until this point, began to argue with him "for being on the wrong side of the issue." The mode decision had been an uninteresting technical issue before, but it became a political concern hashed over in the press for days thereafter. Lord Hailsham, Lord President of the Council and Minister of Science for British Prime Minister Harold Macmillan, who had accompanied Wiesner on the trip, later asked Kennedy on Air Force One how the debate would turn out. The president told him that Wiesner would lose: "Webb's got all the money, and Jerry's only got me." Kennedy was right. Webb lined up political support in Washington for the lunar-orbit rendezvous mode and announced it as a final decision on November 7, 1962. This set the stage for the technological development of hardware to accomplish Apollo.

The American Moon Rocket

Wernher von Braun's rocket team in Huntsville gained distinction in the Moon race by building the biggest, most capable, and most stunningly impressive rocket ever conceptualized. NASA had inherited the effort to develop the Saturn family of boosters used to launch Apollo to the Moon in 1960 when it acquired the Army Ballistic Missile Agency under von Braun. By that time von Braun's engineers were hard at work on the first-generation Saturn launch vehicle, a cluster of eight Redstone boosters around a Jupiter fuel tank. Fueled by a combination of liquid oxygen (LOX) and RP-1 (a version of kerosene), the Saturn I could generate a thrust of 205,000 pounds. This group also worked on a second stage, known in its own right as the Centaur, which used a revolutionary fuel mixture of LOX and liquid hydrogen that could generate a greater ratio of thrust to weight. The fuel choice made this second stage a difficult development effort, because the mixture was highly volatile and could not be readily handled. But the stage could produce an additional 90,000 pounds of thrust. The Saturn I was solely a research-and-development vehicle that would lead toward the accomplishment of Apollo, making ten flights between October 1961 and July 1965. The first four flights tested the first stage, but beginning with the fifth launch the second stage was active, and these missions were used to place scientific payloads and Apollo test capsules into orbit (Table 7).

The next step in Saturn development came with the maturation of the Saturn IB, an upgraded version of the earlier vehicle. With more powerful engines generating 1.6 million pounds of thrust from

Table 7
Saturn Rocket Launches

Date	Program	Launch Vehicle	Launch Results
10/27/1961	Saturn I	SA-1	Successful
4/25/1962	Saturn I	SA-2	Successful
11/16/1962	Saturn I	SA-3	Partially successful
3/28/1963	Saturn I	SA-4	Successful
1/29/1964	Saturn I	SA-5	Successful
5/28/1964	Saturn I	SA-6	Successful
9/18/1964	Saturn I	SA-7	Successful
2/16/1965	Saturn I	SA-9	Successful
5/25/1965	Saturn I	SA-8	Successful
7/30/1965	Saturn I	SA-10	Successful
2/26/1966	Saturn I	AS-201	Partially successful
7/5/1966	Saturn IB	AS-203	Successful
8/25/1966	Saturn I	AS-202	Successful
11/9/1967	Saturn V	*Apollo 4*	Successful
1/22/1968	Saturn IB	*Apollo 5*	Successful
4/4/1968	Saturn V	*Apollo 6*	Partially successful
10/11/1968	Saturn IB	*Apollo 7*	Successful
12/21/1968	Saturn V	*Apollo 8*	Successful
3/3/1969	Saturn V	*Apollo 9*	Successful
5/18/1969	Saturn V	*Apollo 10*	Successful
7/16/1969	Saturn V	*Apollo 11*	Successful
11/14/1969	Saturn V	*Apollo 12*	Successful
4/11/1970	Saturn V	*Apollo 13*	Successful
1/31/1971	Saturn V	*Apollo 14*	Successful
7/26/1971	Saturn V	*Apollo 15*	Successful
4/16/1972	Saturn V	*Apollo 16*	Successful
12/7/1972	Saturn V	*Apollo 17*	Successful
5/23/1973	Saturn IB	SL-2	Successful
7/23/1973	Saturn IB	SL-3	Successful
7/15/1975	Saturn IB	ASTP	Successful

Source: Roger E. Bilstein, *Stages to Saturn: A Technological History of the Apollo/Saturn Launch Vehicles* (Washington, D.C.: NASA SP-4206, 1980), Appendix C, 414–419.

the first stage, the two-stage combination could place 62,000-pound payloads into Earth orbit. The first flight, on July 5, 1966, tested the capability of the booster and the Apollo capsule in a suborbital flight. Eighteen months later, after one flight employing the Saturn I and another introducing the Saturn V, came the January 22, 1968, launch of a Saturn IB with both an Apollo capsule and a lunar-landing module aboard for orbital testing.

The largest launch vehicle of this family, the Saturn V, represented the culmination of those earlier booster development and test programs. Standing 363 feet tall, with three stages, this was the vehicle that could take astronauts to the Moon. The first stage generated 7.5 million pounds of thrust from five massive engines developed for the system. These engines, known as F-1s, were some of the most significant engineering accomplishments of the program, requiring the development of new alloys and different construction techniques to withstand the extreme heat and shock of firing. The thunderous sound of the first static test of this stage, taking place at Huntsville, Alabama, on April 16, 1965, brought home to many that the Kennedy goal was within technological grasp. For others, it signaled the magic of technological effort; one engineer even characterized rocket engine technology as a "black art" without rational principles. The second stage presented enormous challenges to NASA engineers and very nearly caused the lunar-landing goal to be missed. Consisting of five engines burning LOX and liquid hydrogen, this stage could deliver 1 million pounds of thrust. It was always behind schedule, and required constant attention and additional funding to ensure completion by the deadline for a lunar landing. Both the first and third stages of this Saturn vehicle development program moved forward relatively smoothly. (The third stage was an enlarged and improved version of the IB, and had few developmental complications.)

Despite these complications, the biggest problem with Saturn V

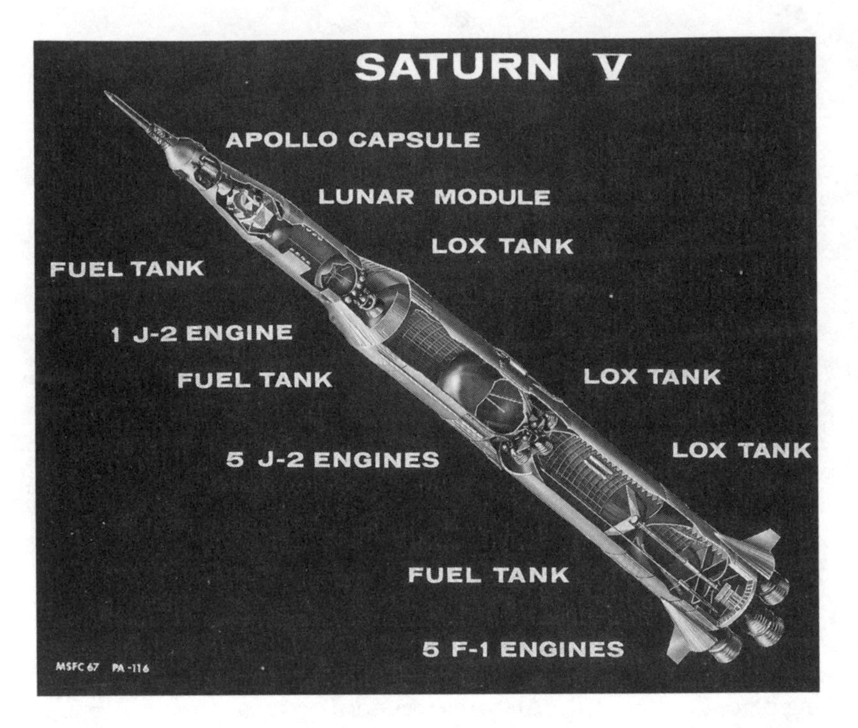

Figure 12. Cutaway illustration of the U.S. Saturn V Moon rocket in 1967 with major components labeled.

lay not with the hardware but with the clash of philosophies toward development and test. The von Braun rocket team had made important technological contributions. Its conservative engineering practices enjoyed popular acclaim. The team tested each component of each system individually, then assembled them for a long series of ground tests. The engineers would then launch each stage individually before assembling the whole system for a long series of flight tests. While this practice ensured thoroughness, it was costly in both money and time, and NASA had neither commodity to spare. George E. Mueller, the head of NASA's Office of Manned Space Flight, disagreed with this approach. Drawing on his experience with the air force and the aerospace industry, and shadowed by the twin bugaboos of schedule and cost, Mueller advocated what he called the "all-up" concept, in which the entire Apollo-Saturn system was tested together in flight, without the laborious preliminaries.

A calculated gamble, the first Saturn V test launch, *Apollo 4*, took place on November 9, 1967, with the entire Apollo-Saturn combination. A second test followed on April 4, 1968, and even though it was only partially successful—the second stage shut off prematurely and the third stage, needed to insert the Apollo payload into lunar trajectory, failed—Mueller declared that the test program had been completed and that the next launch would have astronauts aboard. The gamble paid off. In seventeen test and fifteen piloted launches, the Saturn booster family suffered only one failure, the unmanned *Apollo 6* test, when a stage failed during launch.

The Apollo Spacecraft

Almost with the announcement of the lunar-landing commitment in 1961, NASA technicians began an aggressive program to develop a reasonable configuration for the trip to lunar orbit and back. What they came up with was a three-person command module capable of sustaining human life for two weeks or more in either Earth orbit or a lunar trajectory; a service module holding oxygen, fuel, maneuvering rockets, fuel cells, and other expendable and life-support equipment that could be jettisoned upon reentry to Earth; a retrorocket package attached to the service module for slowing to prepare for reentry; and finally, a launch escape system that was discarded upon achieving orbit. The teardrop shaped command module had two hatches, one on the side for entry and exit of the crew at the beginning and end of the flight and one in the nose with a docking collar for use in moving to and from the lunar-landing vehicle.

Work on the Apollo spacecraft stretched from November 28,

1961, when the prime contract for its development was assigned to North American Aviation, to October 22, 1968, when the last test flight took place. In between there were various efforts to design, build, and test the spacecraft both on the ground and in suborbital and orbital flights. For instance, on May 13, 1964, NASA tested a boilerplate model of the Apollo capsule atop a stubby Little Joe II military booster, and another Apollo capsule achieved orbit on September 18, 1964, when it was launched atop a Saturn I. By the end of 1966 NASA leaders declared the Apollo command module ready for human occupancy.

The NASA project manager for the Apollo spacecraft, Joseph F. Shea, oversaw the spacecraft's design and construction with verve and style, driving hard to meet a standard of excellence not reflected in written documents. With Bob Gilruth's support and encouragement at the Manned Spacecraft Center in Houston, Shea browbeat contractors, other NASA officials, and experts who weighed in on the system. A special target was the rocket team under Wernher von Braun at the Marshall Space Flight Center. Shea specialized in systems engineering and integration, taking a holistic approach that controlled every aspect of the project. His intrusion into decisions viewed by von Braun's rocketeers as within their purview led to more than one flareup that had to be resolved by testy meetings between von Braun and Gilruth. Regardless, as NASA's George Mueller recalled: Shea "contributed a considerable amount of engineering innovation and project management skill." While those working for him enjoyed his eccentricities—especially bad puns and decidedly unprofessional clothing choices—they also recognized his dedication to the effort. He made himself a nuisance, too, often by moving into the construction site and sleeping on a cot during crucial times. By the end of 1966 he believed he had a spacecraft ready for human occupancy, but that was not the case.

The *Apollo 1* Fire

As these development activities were taking place, tragedy struck the Apollo program. On January 27, 1967, Apollo-Saturn (AS) 204, scheduled to be the first spaceflight with astronauts aboard the capsule, was on the launch pad at Kennedy Space Center in Florida, moving through simulation tests. The three astronauts to fly on this mission—Gus Grissom, Edward White, and Roger B. Chaffee—were aboard, running through a mock launch sequence. At 6:31 P.M., after several hours of work, a fire broke out in the spacecraft, fed by the pure oxygen atmosphere intended for the flight. In a flash, flames engulfed the capsule. It took the ground crew five minutes to open the hatch. When they did so, they found three bodies, the astronauts all dead from asphyxiation. Although three other astronauts had been killed before this time—all in plane crashes—these were the first deaths directly attributable to the U.S. space program.

Shock gripped NASA and the nation during the days that followed. James Webb, NASA administrator, told the media at the time, "We've always known that something like this was going to happen soon or later. . . . Who would have thought that the first tragedy would be on the ground?" As the nation mourned, Webb went to President Lyndon Johnson and asked that NASA be allowed to handle the accident investigation and direct the recovery from the accident. He promised to be truthful in assessing blame and pledged to assign it to himself and NASA management as appropriate. The day after the fire NASA appointed an eight-member investigation board, chaired by Floyd L. Thompson, a longtime NASA official and director of the Langley Research Center in Hampton, Virginia. The board set out to discover the details of the tragedy: what happened, why it happened, whether it could happen again, what was at fault, and how could NASA recover? The members of the board learned that the fire had been caused by a short circuit in the electrical system that ignited combustible materials in the spacecraft fed by the oxygen atmosphere. They also found that it could have been prevented and called for several modifications to the spacecraft, including a move to a less oxygen-rich environment. Changes to the capsule followed quickly, and within a little more than a year it was ready for flight.

Webb reported these findings to various congressional committees and took a personal grilling at every meeting. His answers were sometimes evasive and always defensive. The *New York Times*, which was usually critical of Webb, had a field day with this situation and said that NASA stood for "Never a Straight Answer." While the ordeal was personally taxing, whether by happenstance or design Webb deflected much of the backlash over the fire from both NASA as an agency and from the Johnson administration. While he was personally tarred with the disaster, the space agency's image and popular support were largely undamaged. Webb himself never recovered from the stigma of the fire, and when he left NASA in October 1968, even as Apollo was nearing a successful completion, few mourned his departure.

The AS 204 fire also troubled Webb ideologically during the months that followed. He had been a high priest of technocracy ever since coming to NASA in 1961, arguing for the authority of experts, well organized and led, and with sufficient resources to resolve the "many great economic, social, and political problems" that pressed the nation. In *Space Age Management*, published in 1969, he wrote, "Our Society has reached a point where its progress and even its survival increasingly depend upon our ability to organize the complex and to do the unusual." He believed he had achieved that model organization for complex accomplishments at NASA.

Yet that model structure of exemplary management had failed to anticipate and resolve the shortcomings in the Apollo capsule design and had not taken what seemed in retrospect to be normal precautions to ensure the safety of the crew. The system had broken down. As a result Webb became less trusting of other officials at NASA and gathered more and more decision-making authority to himself. This wore on him during the rest of his time as NASA administrator, and the failure of the technological model for solving problems was an important forecaster of a trend in American culture: technology came to be blamed for a good many of society's ills. That problem would be particularly present as NASA tried to win political approval of later NASA projects.

No one took the Apollo accident more personally than Joe Shea and his counterpart at North American Aviation, Harrison Storms. Both were reassigned afterward, Shea in part because of the psychological toll the deaths of the astronauts had on him, as he self-medicated with alcohol and barbiturates. Within a few weeks, Chris Kraft confided that Shea's erratic behavior was hampering the recovery from the fire. He recalled one meeting on the spacecraft, at which "Joe Shea got up and started calmly with a report on the state of the investigation. But within a minute, he was rambling, and in another thirty seconds, he was incoherent. I looked at him and saw my father, in the grip of dementia praecox. It was horrifying and fascinating at the same time." Webb also worried about Shea, and asked him to come to Washington to help the Apollo program from headquarters, promising Shea that he would work only for Webb. He did not tell Shea that he would also have no one working for him. Rocco Petrone, a director of the George C. Marshall Space Flight Center in the 1970s, had been in the blockhouse at the Kennedy Space Center sitting next to astronaut Deke Slayton when the fire broke out. He blamed Shea for the accident, supposedly telling him, "You are a menace and you are to blame for the fire. When you die, I will come and piss on your grave."

Shea left NASA in July 1967 after the Apollo fire, moving to Raytheon, where he worked for many more years. He served as a consultant to NASA in later years but never worked directly for the agency again. Although his replacement was not announced as a punishment, the public interpreted it as such, and had it not been for the accident there is no reason to believe that Shea would have been replaced just as the program neared flight stage. In Storms's case, North American sacrificed him to get back into NASA's good graces and recover from the accident. Storms never forgave NASA for having been forced to "take the fall."

The Lunar Module

If the Saturn launch vehicle and the Apollo spacecraft were difficult technological challenges, the third part of the hardware for the Moon landing, the lunar module (LM), was the mission's problem child. Begun a year later than it should have been, the LM was consistently behind schedule. It represented the most serious design challenge of the whole Moon mission, not least because it was required to safely convey a crew to a soft landing on the surface of another world and later return them to a lunar orbit, where the crew could be reunited with the command module for the trip back to Earth. None of these operations had been done before, so it is perhaps understandable that the LM was consistently behind schedule and over budget.

Much of the project's difficulties turned on the demands of devising two separate spacecraft components—one for descent to the Moon's surface and one for ascent back to the command module. Both engines had to work perfectly to ensure that the astronauts were not left stranded on the lunar surface without any means of getting home. The landing structure likewise presented problems; it had to be light and sturdy and shock resistant. Guidance, maneuverability, and spacecraft control also caused no end of headaches for the LM engineers. After various engineering problems, an ungainly looking vehicle emerged that was finally declared flight-ready in early 1968. It would be piloted by two astronauts standing upright.

Thomas J. Kelly, Grumman's chief designer of the lunar module, remembered what the design had to accomplish: "The command module was totally dominated by the need to reenter Earth's atmosphere, so it had to be dense and aerodynamically streamlined and all that, whereas the lunar module didn't want any of that. It wanted to be able to land on the Moon and operate in an unrestricted environment in space and on the lunar surface. It ultimately resulted in a spindly, gangly-looking, very lightweight vehicle that was just the opposite of all the attributes of the command module. If you tried to do that all in one vehicle, it would be a real problem. I don't know how you would have done it. But this way, with this mission approach, it was very neatly divided in two halves."

Soviet and American Lunar Space Suits

In addition to spacecraft technology, both the Soviet and American Moon programs invested heavily in the development of space suits for their respective lunar missions. The resultant suits were strikingly different. In the mid-1960s Soviet engineers built the Krechet-94 extravehicular activity (EVA) suit. Weighing 198 pounds and worn over a liquid-cooled undergarment, the suit could operate continuously for ten hours, and had a lifespan of forty-eight hours. It also had a semirigid structure, with the hard upper torso section constructed from aluminum, but its most important innovation was the hatch opening at the suit's backpack, which made it easier for cosmonauts to get into and out of it. The Krechet-94's chest-mounted instrument panel offered cosmonauts excellent control over their own life-support systems. While Krechet-94 was never used on the Moon, its design influenced many later Russian and American EVA suits.

The American lunar suit, the A7L, was more complicated. This seventh iterative version of the Apollo suit was designed to provide life support during periods either on the Moon or during weightless EVAs in space. It had a longer operational life than the Soviet version and could withstand as much as 115 hours of use. Like the Krechet-94, A7L also had a liquid-cooled undergarment, but it had an additional two-piece pressure suit and a further one-piece outer coverall. The suit's helmet, gloves, and boots were all attached individually. While the A7L was certainly effective, keeping astronauts safe on their various lunar excursions and other EVAs, it was incredibly laborious to put on and take off, a concern that was factored into the development of later NASA suits.

Soviet Rocket Technology

While the Soviet Union publicly denied throughout the 1960s that it was participating in a space race, it competed aggressively to best the Americans. Khrushchev had set in motion two distinct efforts in the aftermath of JFK's Moon-landing decision, both of them intensely secret. The first, a lunar flyby with cosmonauts using the Soyuz 7K-L1 (Zond) spacecraft launched atop the Proton-K rocket, proceeded under the direction Vladimir Chelomey, whose patronage from Khrushchev brought him favor in acquiring resources for this flight. This program, L3, enjoyed the support of Valentin Glushko, whose rocket engines were the finest available in the Soviet Union. Glushko had no love for Sergei Korolev, whose design bureau competed with Chelomey's efforts. While Korolev scored the mission to land cosmonauts on the Moon, he constantly faced bureaucratic infighting from other design bureaus. This gang of rivals fought to a near standstill for priority in the Soviet system and the resources that would be needed to reach the Moon. If there was one fundamental difference between success as perceived by the Americans and by the Soviets, it was that NASA had a clear and public mandate to reach the Moon by the end of the decade, which no one could challenge, while the Soviets had no one decision-making authority to which all rallied. No decision was ever final for the Soviets, and navigating the politics of the Kremlin was as convoluted as a Dostoevsky novel.

The bureaucratic challenges were laid bare in the aftermath of Korolev's death in January 1966. The management of the Soviet side of the space race was already in disarray, but with Korolev no longer a participant what gravitas he held was gone with him. The Soviet Moon program declined into bickering among various design bureau heads and Kremlin busybodies. Vasily Mishin took over for Korolev, but he was unable to corral the funding necessary to accomplish the purpose. For his part, Chelomey argued that the circumlunar program powered by Glushko's Proton rocket should receive priority over the Korolev/Mishin-led landing effort. This bureaucratic infighting cost the Soviets in terms of time and money. Meantime, design work on the N1 rocket progressed. It stood 345 feet tall and had a mass of about 2,750 tons. The first stage had thirty engines, the second stage had eight, and the third had four. All the engines were powered by liquid oxygen and kerosene, per Korolev's original decision. Lacking time and money, Soviet designers decided to skip full-scale testing of the first stage on the ground. This decision proved fatal, since in four test flights the rocket never did perform successfully.

The upper stage included a translunar booster stage, an upper stage for braking into lunar orbit, a lunar orbiter, and a lunar lander. Unlike the American Apollo spacecraft, which carried three people, the Soviet spacecraft would carry two cosmonauts, one to land on the Moon and one to circle the Moon.

Committed to the end, Mishin planned another spectacular in the game of one-upmanship, seeking to send the first cosmonauts around the Moon in December 1968, a few days before NASA's planned *Apollo 8* Christmastime circumlunar mission with three astronauts. *Zond 6* was an attempt, like the Leonov spacewalk in 1965, to beat a planned American mission to the Moon. Had the Soviets pulled it off, the Kremlin might have extended additional funding for the space race, including a mandate for additional space spectaculars to best the Americans. But when *Zond 6* failed to land safely in a test, there was no choice but to postpone a piloted Zond mission. After *Apollo 8*'s singular success, the Kremlin saw no need to continue the circumlunar program and cancelled it in 1970.

The Soviet Moon-landing project fared no better. After years of delay, the first test of the large N1 Moon rocket in February 1969 also failed. A success might have meant that the Soviet Union could once again one-up the U.S. effort by beating the Americans to a landing. Because of a fire at the bottom of the first stage, however, all engines shut down seventy seconds after launch, and the rocket did not reach orbit. During a second test launch in July 1969, only two weeks before the *Apollo 11* landing mission, the N1 rocket failed to rise beyond 650 feet before collapsing in a massive engine explosion that destroyed the launch complex. As the successful American Moon landings continued, a June 1971 N1 launch also ended in failure; this time the booster's first-stage engines shut down fifty seconds into flight. A final launch in November 1972 ended in a failure of the second stage, with the rocket exploding in flight.

In May 1974 the Kremlin shut the whole program down. Mishin protested, but to no avail, as government officials directed that the remaining hardware be destroyed to validate the Soviets' continuing denial that the USSR had ever intended to engage in a race to the Moon. While a few remaining test articles and models of the lunar orbiter and lander are in Russian museums, few other components of the hardware remain. N1 rocket pieces are on display at the Baikonur Cosmodrome, a grim reminder of what might have been.

Glushko, Korolev's rival, took over as head of the Soviet space program, while Mishin was cashiered after the N1 debacle. Mishin became a nonperson in the space program; his name was never again mentioned in any account of Soviet efforts until the collapse of the USSR and the death of Glushko. Mishin lived until 2001 and told his story effectively in the first decade after the Soviet collapse, based in part on an extensive, fascinating, and revealing set of contemporary diaries that documented just how chaotic the Soviet lunar program had been.

For all of the individual capabilities of those associated with the Soviet lunar program, the dominant reasons for American success and Soviet failure in reaching the Moon involved three fundamental issues:

1. Clear mandates for the Americans and a lack of them for the Soviets, with resulting differences in establishment of priorities and allocation of resources over a lengthy technology development period;
2. Clear lines of authority with a singular decision maker (NASA) for the Americans and an absence of such concentration in the Soviet Union. Korolev held together through force of will early human space-exploration efforts, but with his death no one else had the gravitas to continue;
3. More advanced technical capabilities for the Americans than for the Soviets. For example, Soviet rocket engine technology was incapable of building the large engines developed by the Americans. It took thirty Soviet engines to do the same job as five American engines in launching the Moon rocket. The Soviets never were able to make their engines work well together and reach orbit.

Effectively, the race to the Moon was over before the first Apollo mission ever went there. No one knew that at the time, of course, in either the United States or the USSR, but in retrospect it was objectively true.

SEVEN

Realization

What might we make of the culmination of Apollo, from the first test flights to the various human lunar missions? The realization of this achievement—the individual missions, including direction of the Apollo missions from Mission Control, and launch operations at the Kennedy Space Center—represents one of the high points in the history of the Cold War (Table 8). A reconsideration of the truly remarkable aspect of actually flying to the Moon offers insights into the space race. Looking back fifty years, it also points up a core question about the history of spaceflight, "If we could send a man to the Moon" in 1969, then "why can't we send a man to the Moon" in the early twenty-first century? Examining how the Soviet Union and the rest of the world reacted to the American success opens a door for a reconsideration of the entire episode. The space race led to the desire to change the nature of the competition between the United States and the USSR. Détente offered a way for the two nations to cooperate, and led to the spectacularly successful Apollo-Soyuz Test Project of 1975; the Soviet mission

Table 8

Apollos 7–13

Mission	Launch Date	Crew	Flight Time (days:hrs:mins)	Highlights
Apollo 7	Oct. 11, 1968	Walter M. Schirra, Jr. Donn F. Eisele R. Walter Cunningham	10:20:9	First U.S. 3-person mission
Apollo 8	Dec. 21, 1968	Frank Borman James A. Lovell, Jr. William A. Anders	6:3:1	First human orbit(s) of Moon; first human departure of Earth's sphere of influence; highest speed attained in human flight to date
Apollo 9	Mar. 3, 1969	James A. McDivitt David R. Scott Russell L. Schweickart	10:1:1	Successfully simulated in Earth orbit operation of lunar module to landing and takeoff from lunar surface and rejoining with command module
Apollo 10	May 18, 1969	Thomas P. Stafford John W. Young Eugene A. Cernan	8:0:3	Successfully demonstrated complete system, including lunar module to less than nine miles from the lunar surface
Apollo 11	Jul. 16, 1969	Neil A. Armstrong Michael Collins Buzz Aldrin	8:3:9	First human landing on Moon
Apollo 12	Nov. 14, 1969	Charles Conrad, Jr. Richard F. Gordon, Jr. Alan L. Bean	10:4:36	Second human lunar landing; explored surface of Moon, retrieved parts of *Surveyor 3* spacecraft, which landed in Ocean of Storms on Apr. 19, 1967; demonstrated precision landing
Apollo 13	Apr. 11, 1970	James A. Lovell, Jr. Frederick Haise, Jr. Jack Swigert, Jr.	5:22:54	Aborted landing mission when accident in service module damaged spacecraft; crew successfully returned to Earth

Source: NASA, *Aeronautics and Space Report of the President, 1974 Activities* (Washington, D.C.: NASA, 1975), appendix C, 137–139.

commander was Alexei Leonov, the first human to walk in space in 1965.

Testing the Apollo/Saturn V Hardware

Many people doubted NASA's capabilities in the aftermath of the January 1967 Apollo fire, which killed three astronauts. But that doubt seems to have abated in 1968, and was seemingly swept away altogether during the successful lunar-landing missions between 1969 and 1972. The change began with the October 1967 test flight of *Apollo* 7. Though the *Apollo* 7 mission is overshadowed by later, more ambitious missions, it was an enormous confidence builder, not only for NASA and those working in the space program but also for the public at large. As an Earth orbital shakedown, the *Apollo* 7 mission proved the spaceworthiness of the Apollo spacecraft, but it did much more, reviving the widespread belief that Moon landing was possible, and that NASA could accomplish it.

A sampling of political cartoons from major dailies around the country in late October 1968 finds *Apollo* 7 the subject of sustained praise. One shows a beaming Uncle Sam in a Greek toga reclining on clouds, with the Moon in the background. He holds an Apollo/Saturn stack in his arm as if throwing it as a javelin, and sports a gold medal around his neck. The caption reads, "A Gold Medal in the Lunar Olympics." Another cartoon shows a suspension bridge being built from Earth to the Moon. On the bridge's roadway is the statement, "Apollo 7 Success." The caption reads, "Almost Ready for the Ribbon-Cutting." A third editorial cartoon, titled "Another Winner," shows dice coming up "Apollo Seven" emerging from a shaker labeled "Space Risks." Finally, the *Christian Science Monitor* showed *Apollo* 7 splashing down in the ocean silhouetted against a full Moon in the background with the caption, "Splashdown on the Way Up."

While the *Apollo* 7 mission was significant, the testing of the lunar module represented the most serious challenge. Overcoming the lunar module's budgetary and schedule problems, NASA orbited the first LM on a Saturn V test launch in January 1968 and judged it ready for operation. The lunar module showed its mettle during the first two piloted test flights. During *Apollo 9*, March 3–13, 1969, the crew tested the LM in Earth orbit; and on *Apollo 10*, May 18–26, 1969, the LM performed well in lunar orbit, going as close to the surface as 60.9 nautical miles by 8.5 nautical miles. With those successes, the landing phase of the Apollo program commenced.

Apollo 8: The First Trip Around the Moon

Apollo 8 began as another test flight of the Apollo system in Earth orbit, but it turned into arguably the most important mission flown during Apollo other than the *Apollo 11* landing. It took off atop a Saturn V booster from the Kennedy Space Center with three astronauts aboard—Frank Borman, James A. Lovell, Jr., and William A. Anders—for a historic mission to orbit the Moon. Bob Gilruth's senior engineer George M. Low, who had replaced Joe Shea as Apollo spacecraft program manager, proposed turning *Apollo 8* into a circumlunar flight. He recognized that other hardware, especially the lunar module, was not yet ready for testing and that the Soviet Union seemed to be focused on its Zond circumlunar flight in the fall of 1968. Low persuaded Samuel C. Phillips, Apollo program manager at NASA headquarters, to support the scheme, and the two of them gained approval to make *Apollo 8* a circumlunar flight.

The advantages of doing this could be important, both in technical and scientific knowledge gained and as a public demonstration of what the United States could achieve. So far Apollo had been all promise; now the delivery was about to begin. In the summer of 1968 Low broached the idea to Phillips, who carried it to the administrator. After reviews in November the agency reconfigured the mission for a lunar trip.

Apollo 8 launched on December 21, 1968, and after it made one and a half Earth orbits, its third stage began a burn to put the spacecraft on a lunar trajectory. As it traveled outward, the crew focused a portable television camera on Earth, and for the first time humanity saw its home from afar, a tiny, lovely, fragile "blue marble" hanging in the blackness of space. When the craft arrived at the Moon on Christmas Eve, this image of Earth was even more strongly reinforced, when the crew transmitted images of the planet while reading from the chapter of Genesis: "God created the heavens and the Earth, and the Earth was without form and void." The next day the crew fired the boosters for a return flight, "splashing down" in the Pacific Ocean on December 27.

This first circumlunar flight may be best remembered for two noteworthy results. First, the decision to send a Saturn V on only its first flight with a crew around the Moon was one of the gutsiest calls in NASA's history. After the *Apollo 1* fatal capsule fire in January 1967, NASA fell behind schedule to land on the Moon by the end of the decade. To catch up, NASA leaders concocted a bravura scheme to regain momentum by making *Apollo 8* a lunar flyby. The advantages were obvious: NASA acquired the knowledge needed to undertake the landing, and the flight served as a worldwide demonstration of U.S. technological excellence.

Second, during *Apollo 8* Bill Anders from lunar orbit took one of the most critically significant images of the twentieth century, later dubbed *Earthrise*. Usually depicted horizontally, this photograph showed the Moon, gray and lifeless in the foreground, with Earth awash in color and hanging in the blackness of space. *Earthrise* symbolized for more than a generation an emerging environmental consciousness. The astronauts did not realize the impact of their image until well after the fact; it has been lauded by poets and pundits, preservationists and potentates. At a fundamental level *Earthrise* became the ultimate reconnaissance photograph of Earth taken

Figure 13. *Earthrise,* one of the most powerful and iconic images from the Apollo program, was taken in December 1968 during the *Apollo 8* mission. This view of the rising Earth greeted the *Apollo 8* astronauts as they came from behind the Moon after the first lunar orbit. Used as a symbol of the planet's fragility, it juxtaposes the gray, lifeless Moon in the foreground with the blue and white Earth teeming with life hanging in the blackness of space.

from afar. It has found use in all manner of settings and places since it first entered the public's consciousness in December 1968.

Apollo 8 was an enormously significant accomplishment, coming at a time when American society was in crisis. Without question, 1968 had been one of the most tumultuous years in American history. The year's upheaval began on January 23, with North Korea's capture of the American surveillance ship USS *Pueblo*, beginning an eleven-month hostage crisis. A week later the Tet offensive began in Vietnam, with North Vietnamese and Viet Cong troops attacking

Saigon and more than one hundred other cities. In April, the assassination of Martin Luther King, Jr., by a white supremacist led to riots in more than a dozen major cities. Presidential candidate Robert F. Kennedy died in June from a gunman's bullet after winning the California Democratic primary. Summer ended in violence at the Democratic National Convention in Chicago, and the troubles of the time were further demonstrated in October when two African-American medalists at the Olympic Games raised their fists in silent protest to the injustice they saw around them. A sizable percentage of the U.S. population believed the election of Richard M. Nixon as president in November signaled a national torpor at best and a distinct turn toward oblivion at worst. Emotions ran high, and while interpretations varied, all agreed that the United States was seemingly unraveling.

At the end of 1968, *Apollo 8*'s success felt like a salve on an open wound for many Americans. Frank Borman has often told the story of how he received a telegram after the mission that read: "Thank you, Apollo 8. You saved 1968."

Apollo 11: The Main Event

Then came the big event. Astronauts Neil A. Armstrong, Buzz Aldrin, and Michael Collins were thoroughly prepared and well rehearsed for their mission. NASA scientists had found a suitable landing site that was geologically interesting but also an open flat plain. Engineers had drilled the crew on every aspect of the mission's week-long flight. They practiced space walks in a deep swimming pool dubbed the Neutral Buoyancy Simulator, donned and doffed spacesuits repeatedly, and trained for entry and exit of both their capsule and the lunar module. In addition to all this, Armstrong practiced landing the LM in the Moon's lower gravity, roughly one-sixth of Earth, using the Lunar Landing Research Vehicle simulator for practice in Houston.

On July 16, 1969, *Apollo 11* launched from the Kennedy Space Center without incident and began the three-day trip to the Moon. On July 20, the LM, dubbed *Eagle* and crewed by Armstrong and Aldrin, separated from the command/service module (CSM) to begin its descent toward the lunar surface. The landing was difficult. As the LM neared the surface, Armstrong realized that the automatic landing system was poised to set them down in the middle of a boulder field, so he took manual control and searched for another landing spot. As he slowed the descent over the lunar surface, the LM used more and more of its fuel, setting off low-fuel alarms in the spacecraft. Aldrin called out the altitude and the status of fuel. With just eleven seconds of fuel left, amid rising tension among those at Mission Control, Armstrong finally set *Eagle* down on the lunar surface, announcing, "Contact light. Houston, Tranquility Base here. The Eagle has landed." Charlie Duke, the astronaut at Mission Control responsible for communicating with the mission crew, responded in a flustered voice, "Roger, Tranquility, we copy you on the ground. You got a bunch of guys about to turn blue. We're breathing again. Thanks a lot!"

After the landing the crew had been scheduled to sleep for five hours, but Armstrong and Aldrin elected to skip it, reasoning that they would be too excited to sleep. After some final checks the pair suited up. Armstrong left *Eagle* to set foot on the Moon, telling millions on Earth that it was "one small step for man—one giant leap for mankind." (Armstrong later added "a" when referring to "one small step for a man" to clarify the first sentence delivered from the Moon's surface.) Aldrin soon followed, and while he was not first on the Moon, he paused for a moment to relieve himself just as he reached the surface, claiming a different type of first for himself. Thereafter, the two lumbered around the landing site in their bulky space suits in the lunar gravity, one-sixth that of Earth, planting an American flag, collecting soil and rock samples, and setting up scientific experiments. The ceremonial planting of the flag, conjuring images of Europeans doing the same in others parts of the world from the sixteenth to the nineteenth centuries, notably did not include a claim of the Moon for the United States. Instead, the astronauts proclaimed that they "came in peace for all mankind." That phrase was not accidental, and its meaning was not lost on the rest of the world. The next day the crew launched the LM to rendezvous with the Apollo CSM orbiting overhead before returning to Earth.

Some were concerned that the crew might be stranded on the Moon. Shortly before the flight astronaut Bill Anders contacted the White House to warn that they should be ready in the event of an accident. Richard Nixon's speechwriter, William Safire, prepared an eloquent speech should it be necessary. Nixon would have declared: "They will be mourned by their families and friends; they will be mourned by their nation; they will be mourned by the people of the world; they will be mourned by a Mother Earth that dared send two of her sons into the unknown." It concluded: "Others will follow, and surely find their way home. Man's search will not be denied. But these men were the first, and they will remain the foremost in our hearts." After *Apollo 11*, Safire filed the speech way, transferred his papers to the National Archives, and forgot about it. Researchers rediscovered it in 2009, reminding all how risky space exploration was. Fortunately, the speech proved unnecessary.

Apollo 11 also unified for a brief moment an American nation divided by political, social, racial, and economic tensions in the summer of 1969. Virtually everyone old enough recalls where they were when *Apollo 11* touched down on the Moon. One seven-year-old boy from San Juan, Puerto Rico, said of the first Moon landing: "I kept racing between the TV and the balcony and looking at the Moon to see if I could see them on the Moon." He recalled never

Figure 14. Aldrin at the flag is an iconic image from *Apollo 11*. This image also circled the globe immediately after its release in July 1969 and has been used for all manner of purposes since that time, including for the logo for MTV in its early years. The flag in this image proved a powerful trope of American exceptionalism. It also has often been used by Moon-landing deniers as evidence that the landing was filmed on Earth, because the flag appears to be waving in the breeze, and we all know there is no breeze on the Moon. When astronauts were planting the flagpole, they rotated it back and forth to better penetrate the lunar soil (a maneuver familiar to anyone who has set a blunt tent post). Of course, the flag waved—no breeze required!

being more proud as an American citizen than in the summer of 1969.

The Reverend Ralph Abernathy, successor to Martin Luther King, Jr., as head of the Southern Christian Leadership Conference, led a protest to the *Apollo 11* launch to call attention to the plight of the poor of the United States. The protesters held an all-night vigil as the countdown proceeded, then marched with two mule-drawn wagons as a reminder that while the nation spent significant money on the Apollo program, poverty ravaged many Americans' lives. As Hosea Williams said at the time, "We do not oppose the Moon shot. Our purpose is to protest America's inability to choose human priorities."

This protest pointed up the confluence of high technology challenges and the more mundane but ever-present problems of American society. Abernathy asked to meet with the NASA leadership, and the space agency's administrator, Thomas O. Paine, met with Abernathy before the launch. Paine recorded the incident:

> We were coatless, standing under a cloudy sky, with distant thunder rumbling, and a very light mist of rain occasionally falling. After a good deal of chanting, oratory and lining up, the group marched slowly toward us, singing "We Shall Overcome." In the lead were several mules being led by the Rev. Abernathy, Hosea Williams and other leading members of the Southern Christian Leadership Conference. The leaders came up to us and halted, facing Julian [Scheer, NASA press secretary] and myself, while the remainder of the group walked around and surrounded us. . . . One fifth of the population lacks adequate food, clothing, shelter and medical care, [Rev. Abernathy] said. The money for the space program, he stated, should be spent to feed the hungry, clothe the naked, tend the sick, and house the shelterless.

Abernathy said that he had three requests for NASA: that ten families of his group be allowed to view the launch, that NASA "support the movement to combat the nation's poverty, hunger and other social problems," and that NASA technical people work "to tackle the problem of hunger."

Paine responded by inviting Abernathy and a busload of his supporters to view the *Apollo 11* launch from the VIP site with other dignitaries. Paine told Abernathy that it was difficult to apply NASA's scientific and technological knowledge to the problems of society. "I stated that if we could solve the problems of poverty in the United States by not pushing the button to launch men to the moon tomorrow," Paine said, "then we would not push that button." He added:

> I said that the great technological advances of NASA were child's play compared to the tremendously difficult human problems with which he and his people were concerned. I said that he should regard the space program, however, as an encouraging demonstration of what the American people could accomplish when they had vision, leadership and adequate resources of competent people and money to overcome obstacles. I said I hoped that he would hitch his wagons to our rocket, using the space program as a spur to the nation to tackle problems boldly in other areas, and using NASA's space successes as a yardstick by which progress in other areas should be measured. I said that although I could not promise early results, I would certainly do everything in my own personal power to help him in his fight for better conditions for all Americans, and that his request that science and engineering assist in this task was a sound one which, in the long run, would indeed help.

Paine then asked Abernathy, who had scheduled a prayer meeting later that day with his protestors, that they "pray for the safety of our astronauts." As Paine recalled, "He responded with emotion that they would certainly pray for the safety and success of the astronauts, and that as Americans they were as proud of our space achievements as anybody in the country."

Paine realized that the social problems of the United States could not be solved entirely by revectoring resources from NASA to other initiatives. He also agreed that the problems of society were much more complex and defied resolution using the tools, knowledge, and resources employed to accomplish Project Apollo. While it might be tempting to generalize from the experience of NASA during the 1960s that its success might be duplicated elsewhere, such was not the case.

Apollo 11 inspired an ecstatic reaction around the globe, as everyone shared in the success of the mission. Ticker-tape parades, speaking engagements, public relations events, and a world tour by the astronauts served to create goodwill both in the United States and abroad. It became much more than an American achievement; it was a "human" triumph. Newspapers' front pages virtually everywhere displayed this mood. NASA estimated that because of nearly worldwide radio and television coverage, more than half of the planet's population had followed *Apollo 11*.

One of the objectives of the Apollo program had been to demonstrate American technological competence, thereby bringing allies to the Western Cold War bloc. It was successful in achieving this objective. Although the Soviet Union tried to jam *Voice of America* radio broadcasts, most living there and in other Eastern Bloc countries followed the adventure carefully. Police reports noted that streets in many cities were eerily quiet during the Moon walk, as residents watched television coverage in homes, bars, and other public places. U.S. ambassador to Morocco Henry Tasca recalled that *Apollo 11* proved "a unifying experience," captivating everyone from the king to "the street beggar." There could be no doubt, he commented, "that the international position of the United States in all its aspects has been deeply, . . . irreversibly changed" with the Moon landing.

Official congratulations poured in to the U.S. president from other heads of state, even as informal ones went to NASA and the astronauts. All nations having regular diplomatic relations with the United States sent their best wishes in recognition of the success of the mission. The People's Republic of China, with no such official relations, made no formal statement to the United States on the *Apollo 11* flight, and the mission was reported only sporadically by news media because Mao Zedong refused to publicize successes by Cold War rivals. Only in February 1972, when Nixon flew to China and met with Mao, did the United States establish formal diplomatic relations with the nation. China now seeks to go to the Moon, and fully recognizes the success of the Apollo program.

The day after the Moon walk, the astronauts launched back to the Apollo capsule orbiting overhead and began the return trip to Earth, splashing down in the Pacific on July 24. This flight rekindled the excitement felt in the early 1960s with John Glenn and the Mercury astronauts. When meeting the crew, President Nixon told a worldwide television and radio audience that the flight of *Apollo 11* represented the most significant week in the history of Earth since the creation. This was political hyperbole, but the successful Moon landing captured the world's attention. As the legendary American journalist Walter Cronkite commented in 2003, those who witnessed the Apollo Moon landing were part of "the lucky generation." They participated in the experiences as humanity "first broke our earthly bonds and ventured into space. From our descendants' perches on other planets or distant space cities, they will look back at our achievement with wonder at our courage and audacity and with appreciation at our accomplishments, which assured the future in which they live."

Apollo 12: Precision Landing

Apollo 12 entered the public's perception in a way different from the first Moon landing. It represented exciting and significant advances in capability over *Apollo 11*, but it did not have a broadcast component, since the camera failed on the lunar surface. Still, the second lunar landing, on November 14–24, 1969, broke open the space race with the Soviet Union. *Apollo 11* might have been a fluke, but this mission required a precision landing close to a robotic Surveyor 3 spacecraft that had touched down on the lunar surface in 1967. If the Americans could pull that off, Soviet space-exploration officials knew, the USSR could not compete in the Moon race. The descent was automatic, with only a few manual corrections by commander Charles "Pete" Conrad and Alan L. Bean. The landing in the southeastern part of the Ocean of Storms brought the lunar module *Intrepid* within walking distance—600 feet—of Surveyor 3.

Conrad's statement when stepping onto the Moon was more spontaneous if less memorable than Neil Armstrong's "one giant leap." "Whoopee!" he said. "Man, that may have been a small one for Neil, but that's a long one for me." Conrad and Bean brought pieces of the Surveyor 3 back to Earth for analysis, and took two Moon walks lasting just under four hours each.

They collected rocks and set up experiments that measured the Moon's seismic activity, solar wind flux, and magnetic field. Meanwhile Richard Gordon, on board the *Yankee Clipper* in lunar orbit, took multispectral photographs of the surface. The crew stayed an extra day in lunar orbit taking photographs. When the lunar module ascent stage was dropped onto the Moon after Conrad and Bean rejoined Gordon in orbit, the seismometers the astronauts had left on the lunar surface registered the vibrations, allowing the measurement of seismic activity on the Moon. This experiment, and others of a similar nature, allowed scientists on Earth to measure the density and materials making up the Moon, thereby helping to answer one of the central questions in lunar science: how did the Moon come to be? It took more than a decade for a consensus to emerge that the Moon came about through Earth's collision with another planet soon after the solar system formed, but the measurement of lunar seismic activity was critical to this scientific finding.

One of the most unusual stories of *Apollo 12* involved the Surveyor 3 camera that had journeyed to the Moon on April 20, 1967, and had sat exposed on the lunar surface for thirty-one months before the *Apollo 12* astronauts retrieved it in December 1969. Humans, of course, cannot survive more than a few seconds while exposed to the vacuum of space. Apparently, not so some extremophiles from this planet. After the *Apollo 12* mission, while examining the Surveyor 3 camera, scientists saw evidence of micrometeoroid bombardment. But they also found terrestrial bacteria—*Streptococcus mitis*—that apparently had survived for more than two and a half years in the vacuum of space.

The bacteria occupied only one of thirty-three samples from various parts of the spacecraft, and the question arose whether it predated *Apollo 12*'s visit or resulted from accidental contamination following return from the Moon. No one knows the answer, but there is firm evidence from other space projects to suggest that some forms of microbial life can go into hibernation while in space and revive once they reach a hospitable environment. *Apollo 12*'s mission offered the first inkling that this might be the case. In any case, the low-key response to the finding was amazing to nonscientist Pete Conrad, who recalled in 1991, "I always thought the most significant thing that we ever found on the whole goddamn Moon was that little bacteria who came back and lived and nobody ever said shit about it."

Apollo 13: A Successful Failure

In spite of the success of the other missions to the Moon, only *Apollo 13*, launched on April 11, 1970, came close to matching earlier popular interest. It would have been the third landing, and astronauts James A. Lovell, Jr., Fred W. Haise, Jr., and John L. Swigert, Jr., practiced enthusiastically to achieve important scientific discoveries. *Apollos 11* and *12* had been largely about reaching the surface safely and collecting lunar samples, without much emphasis on geological questions. Lovell even used the motto *Ex Luna, Scientia* to emphasize this objective. They would explore the Fra Mauro Highlands, a mountainous area thought to be formed by ejecta from a crater.

Apollo 13 never reached Fra Mauro. After only fifty-six hours of flight, an oxygen tank in the Apollo service module ruptured and damaged several of the power, electrical, and life-support systems. NASA engineers quickly determined that the spacecraft was dying, turning the lunar module—a self-contained landing craft unaffected by the accident—into a "lifeboat" to provide austere life support for the return trip. In the feature film *Apollo 13* (1995), NASA flight director Gene Kranz announces, "Failure is not an option." While Kranz never actually said that line, he has recalled that he wished he had thought of it at the time, since it fit so well with what NASA did during the crisis. Bringing the crew home alive was now the only objective.

To achieve that goal, the crew and those on the ground fell back on their training, improvised where necessary, and showed their mettle in systematically tackling every problem thrown in the way of a safe return. People around the world watched and waited and hoped as NASA worked to return the crew to Earth. It was a close call, but they returned safely on April 17. More than any other incident in the history of spaceflight, recovery from this accident solidified the world's belief in NASA's capabilities. One might best refer to *Apollo 13* as a "successful failure."

Because of this failure, however, such NASA leaders as Bob Gilruth pulled back from an aggressive spaceflight program to the Moon. *Apollo 13* served as an object lesson of the risks involved in the lunar voyages. It was only a matter of time, Gilruth realized, before astronauts would die in this endeavor. "There were some people who wanted to keep on flying those things, you know, a lot more of them," Gilruth said in 1987. "I said, 'Not me, you get another boy. . . . I'm not going to stay around for it if you're going to keep doing it.'" In fact, Gilruth had always been cautious in space endeavors. He recalled in 1975: "You don't take any risks in this business that you don't have to take." Now that NASA had achieved Kennedy's objectives of landing Americans on the Moon by the end of the decade and demonstrating U.S. preeminence in space, Gilruth argued for a termination of the Apollo landing program. He was not alone, and after a series of more expansive scientific missions between 1970 and 1972, the program ended.

Apollos 14–17: A Scientific Harvest

Upon recovering from the failure of *Apollo 13*, NASA carried out four additional landing missions to the Moon (Table 9). *Apollos 14–17* reaped a harvest of scientific knowledge about the Moon's origin and evolution. The last three missions, beginning with *Apollo 15*, also used a lunar roving vehicle (LRV) to travel farther and stay longer than ever before. This meant that the most interesting surface features, mountains and rilles (long, narrow depressions in the Moon

Table 9

Apollos 14–17

Mission	Launch Date	Crew	Flight Time (days:hrs:mins)	Highlights
Apollo 14	Jan. 31, 1971	Alan B. Shepard Stuart A. Roosa Edgar D. Mitchell	9:0:2	Third human lunar landing; mission demonstrated pinpoint landing capability and continued human exploration using a pull cart
Apollo 15	Jul. 26, 1971	David R. Scott Alfred M. Worden James B. Irwin	12:7:12	Fourth human lunar landing and first Apollo "J" series mission, which carried lunar roving vehicle; Worden's in-flight EVA of 38 minutes, 12 seconds was performed during return trip
Apollo 16	Apr. 16, 1972	John W. Young Charles M. Duke, Jr. T. K. Mattingly II	11:1:51	Fifth human lunar landing, and second with lunar roving vehicle
Apollo 17	Dec. 7, 1972	Eugene A. Cernan Harrison H. Schmitt Ronald E. Evans	12:13:52	Sixth and final Apollo human lunar landing, again with roving vehicle

Source: NASA, *Aeronautics and Space Report of the President, 1974 Activities* (Washington, D.C.: NASA, 1975), appendix C, 137–139.

that resemble channels), were more accessible than if astronauts had to walk there.

Scientists exploited the advancing capabilities of the Apollo missions, asking the astronauts to journey farther from their landing site, placing more than fifty experiments on the surface during various Apollo missions, and training the astronauts as more than amateur geologists. The most important aspect of this was the development of the Apollo Lunar Surface Experiments Package, a set of instruments deployed by the astronauts on the surface to measure such aspects of the environment as soil mechanics, meteoroids, seismic activity, heat flow, lunar ranging, magnetic fields, and solar wind. These science packages deployed on the Moon included experiments that yielded more than ten thousand scientific papers and helped establish a major reinterpretation of the origins and evolution of the Moon.

The scientific community worked with the Apollo astronauts to prepare them for geological fieldwork on the Moon. Never happy that academically trained geologists were rare in the astronaut corps—the only Moon-walking geologist was Harrison Schmitt who had earned a Ph.D. in science from Harvard University—NASA scientists nonetheless worked hard to ensure that the flight crews had the knowledge necessary to undertake useful work on the lunar surface. To a surprising degree they succeeded. Between 1964 and the times of the various missions, the crews undertook classroom study and fieldwork in a variety of settings to prepare for their time on the lunar surface. The flight crews underwent formal education roughly equivalent to a master's degree in geology.

Indicative of the approach taken by some of the astronauts was that of David Scott on *Apollo 15*. He enthusiastically trained for the mission, and once on the Moon concentrated on the scientific efforts. As he recalled: "Most of my thoughts on the Moon were of the geology involved. Our mission was especially heavy in science, trying to understand the geology of the local site and the Apennines—why things occurred as they did."

Scott and crewmate Jim Irwin found the so-called Genesis Rock of anorthosite, more than 4 billion years old, formed in the early stages of the history of the solar system and therefore a window into the origins of the Moon, Earth, and this solar system.

The astronauts on the six missions that landed on the Moon returned almost nine hundred pounds of lunar samples. Since the Moon landings more than sixty research laboratories around the world have continued studies on the Apollo lunar samples. Many analytical technologies, including some that did not exist in 1969–1972 during the Apollo missions, have been used by later scientists.

So What?

In retrospect we can't help being impressed by the efforts made to reach the Moon in the 1960s and reveling in the success of humankind in achieving this striking accomplishment. A spring 1999 poll of opinion leaders sponsored by major news organizations in the United States, for example, ranked the one hundred most significant news events of the twentieth century. The Moon landing came in a close second to the splitting of the atom and its use during World War II. Some respondents found it difficult to choose between the various events. "It was agonizing," CNN anchor and senior correspondent Judy Woodruff said of the selection process. Historian Arthur M. Schlesinger, Jr., summarized the position of many opinion leaders: "The one thing for which this century will be remembered 500 years from now was: This was the century when we began the exploration of space," he commented. Schlesinger said he looked forward toward a positive future and that prompted him to rank the lunar landing first. "I put DNA and penicillin and the computer and the microchip in the first 10 because they've transformed civilization. Wars vanish," Schlesinger said. "Pearl Harbor will be as remote as the War of the Roses," he said, referring to the English civil wars of the fifteenth century. He added, "The order is essentially very artificial and fictitious. It's very hard to decide the atomic bomb is more important than getting on the Moon."

At the same time, the space race was largely a space spectacular, and enthusiasm for expansive public programs to go beyond Earth was not to be repeated. In such a situation, Apollo appears more as something Americans did once upon a time for reasons that have receded far into the background. While many viewed the astronauts and cosmonauts who undertook space missions as explorers akin to fifteenth-century seafarers like Christopher Columbus, the vanguards of sustained human exploration and migration, regard for their accomplishments seems to diminish with every passing year. Early feats of human exploration were entertaining, but the excitement was rooted in a relative lack of understanding about the nature of the Moon and planets, and a view of colonization outmoded even at that time.

The success of Apollo, in retrospect, has facilitated a misperception about the space race and its support by the American public at the time. While there may be reasons to accept that Apollo was transcendentally important at some sublime level, assuming a generally rosy public acceptance of it is at best a simplistic and ultimately unsatisfactory conclusion. Indeed, the public's support for space funding has remained remarkably stable. For example, in the summer of 1965 one-third of the U.S. citizenry favored cutting the space budget, while only 16 percent wanted to increase it. Over the next three and one half years, the number in favor of cutting space spending went up to 40 percent, with those preferring an increase dropping to 14 percent.

In October 1965, a Harris Poll found that several other public issues were nearly as high public priorities as efforts in outer space (Graph 3). Polls in the 1960s also consistently ranked spaceflight

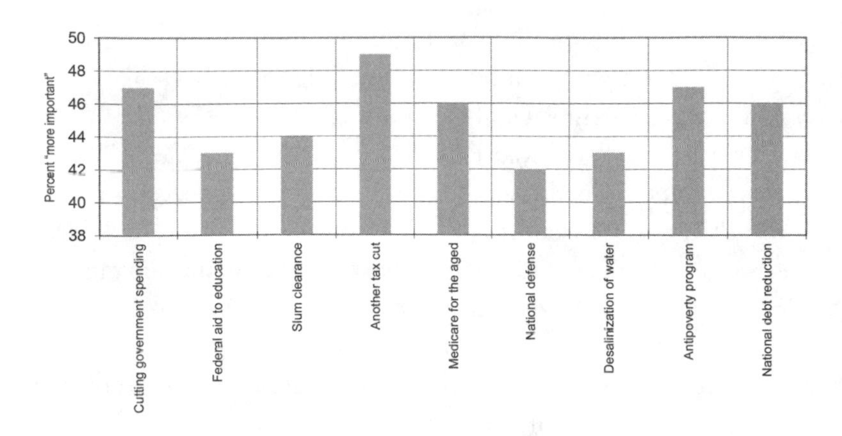

Graph 3. Public Assessment of Costs of
Space Exploration Relative to Other Programs
Source: Harris Poll, 1965. Question: If you had to choose, do you think it more important or less important to spend $4 billion a year on the space program than to spend it on . . . ?

near the top of those programs to be cut in the federal budget. Most Americans seemingly preferred doing something about air and water pollution, job training for unskilled workers, national beautification, and poverty before spending federal funds on human spaceflight. In 1965 *Newsweek* echoed the *Times* story, stating: "The U.S. space program is in decline. The Vietnam war and the desperate conditions of the nation's poor and its cities—which make space flight seem, in comparison, like an embarrassing national self-indulgence—have combined to drag down a program where the sky was no longer the limit." Moreover, throughout most of the 1960s the American public answered the question "Should the Government Fund Human Trips to the Moon?" with a less than positive response (Graph 4).

These statistics do not demonstrate an unqualified support for NASA's effort to race the Soviets to the Moon in the 1960s. They suggest, instead, that the political crisis that brought public support to the initial lunar-landing decision was fleeting and that the public

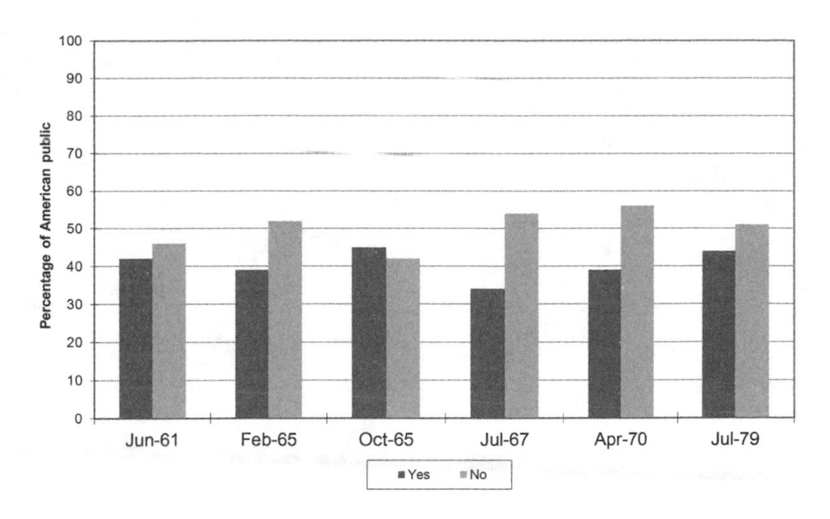

Graph 4. Should the Government Fund Human Trips to the Moon?
Sources: Gallup, Harris, NBC/Associated Press, CBS/New York Times Polls; wording of questions differed slightly.

was never enthusiastic about human lunar exploration, and especially about the costs associated with it. What enthusiasm it may have enjoyed waned over time, until by the end of the Apollo program in December 1972, the program seemed akin to a limping marathoner straining with every muscle to reach the finish line before collapsing.

EIGHT
Revelations

A series of important outcomes resulted from the race to the Moon. For example, by sheer serendipity Apollo taught humanity about itself, and in the process altered our perception of the world on which we live. The cosmonauts, while not getting into translunar space, expressed similar thoughts about the meaning of the first years of the space age and seeing Earth in a new way. The modern environmental movement was galvanized in part by this new perception of the planet and the need to protect it and the life that it supports. Additionally, while science was not the driver behind the race to the Moon, quite a lot of scientific understanding resulted from it. Both the United States and the USSR contributed to this understanding, with robotic probes to the Moon as well as the Apollo landings. Finally, the management of large-scale technological endeavors was a major result of these efforts in both nations.

A New Environmentalism

Every astronaut and cosmonaut commented that seeing Earth from space could not help but change one's perspective. Cosmonaut Yuri

Gagarin said after returning to Earth in 1961, "Orbiting Earth in the spaceship, I saw how beautiful our planet is. People, let us preserve and increase this beauty, not destroy it!" The second cosmonaut, Gherman Titov, also commented on the beauty of Earth. "It's a pity I flew only once," he commented. "A space flight is like a drug—once you experience it, you can't think of anything else." John Glenn flew the first handheld camera into space in 1962, and his pictures of Earth quickly became the most memorable aspect of his flight. Couple this with the emerging awareness of Earth as a fragile habitat that must be preserved, as articulated in such books as Rachel Carson's *Silent Spring* (1962), and a global modern environmental movement gained momentum in the 1960s.

Apollo 8 was critical to this fundamental change, as it treated the world to the first pictures of Earth from afar. Writer Archibald MacLeish summed up the feelings of many people when he wrote at the time of Apollo, "To see the Earth as it truly is, small and blue and beautiful in that eternal silence where it floats, is to see ourselves as riders on the Earth together, brothers on that bright loveliness in the eternal cold—brothers who know now that they are truly brothers."

Earthrise certainly made a difference, but so did the image of Earth from *Apollo 17*. As early as 1966, American environmental activist Stewart Brand had begun a campaign for NASA to release an image of the whole Earth in space. Brand even made up buttons that asked, "Why haven't we seen a photograph of the Whole Earth yet?" He sold them on college campuses and mailed them to prominent scientists, futurists, and legislators. Not until the *Apollo 17* mission in 1972, however, did "Whole Earth" become a reality. As Brand recalled:

> I turned on my blanket there on the gravel rooftop and looked clear around, it was indeed a circle, a mandala—a nice, finite,

> entire, low-altitude view of the Earth. . . . I just sat all afternoon and tried to think of how we could possibly get a photograph of the whole Earth—that is, of the planet from space. I was a big fan of NASA and of then ten years of space exploration that had gone up to that point, and there we were in 1966, having seen a lot of the moon and a lot of hunks of the Earth, but never the complete mandala. . . . It was a bit odd that for ten years, with all the photographic apparatus in the world, we hadn't turned the cameras that 180 degrees to look back.

To capture this iconic image the astronauts on *Apollo 17* used a 70-millimeter Hasselblad camera, and no one is quite sure who snapped the iconic image. Many credit Harrison Schmitt with taking the photo, but it cannot be determined for certain. Since no humans have made a lunar trip since *Apollo 17*, this was the last opportunity. Stewart Brand put the photograph on the cover of his *Whole Earth Catalog*, first published in 1968. This image, and the other stunning photographs of Earth taken from space, inspired a reconsideration of our place in the universe. It became the iconic image for environmental activists, politicians, and scientists during the annual Earth Day celebrations. They used it as an object lesson of Earth as a small, vulnerable, lonely, and fragile body teeming with life in a dull, black, lifeless void. The planet was self-regulating and ancient, they observed, but humanity was a threat to this place. Earth now required human protection. Astronaut Joseph P. Allen said it best: "With all the arguments, pro and con, for going to the Moon, no one suggested that we should do it to look at the Earth. But that may in fact be the one important reason."

Technological Virtuosity

The space race represented a triumph of engineering and technological virtuosity. For both the Americans, whose successes and failures were on display for all to see, and the Soviets, who hid their failures, it symbolized modernity and forward looking. The message came through loudly around the globe. Astronauts standing on the Moon, cosmonauts engaging in a series of one-upmanship spectaculars, and both nations' ballyhooing their technological virtuosity served well the Cold War competition between the Americans and the Russians. Moreover, there is no question that the successes in space during the 1960s helped to create a culture of competence for these rivals in the Cold War, swaying various other nations to ally with them in the larger geopolitical struggle. It was, like so much of the engagement of Soviet and American foreign policy during the Cold War, war by another means in which no one would be killed, at least not intentionally.

This culture of competence rubbed off on NASA and the Soviet space program as well. For NASA it translated into a level of confidence in American capability, and especially in the ability of government to perform effectively, to resolve problems. Recollections of the Apollo program's technology led many to express wonder at the sophistication of the technical competence that made the Moon landings possible and the genius of those that built the rockets and spacecraft that carried Americans into space. Egyptian-American Farouk El-Baz, a scientist who worked on the program, expressed well this sense of awe at the Moon landings: "Oh, the Apollo program! It was a unique effort all together. When I think about it some 40 years later, I still look at that time with wonder."

This is the case in no small measure because of the relative lack of complexity of the technology used to go to the Moon in the 1960s. Many express wonder that there is more computing power in a pocket calculator than in the Apollo guidance computer. Others are surprised that something as simple as writing in space required the development of a new type of pen, with the ink under pressure so that it could write in a weightless environment.

Figure 15. An artist's concept of the Apollo-Soyuz Test Project (ASTP), the first international docking of the U.S. Apollo spacecraft with the USSR's Soyuz spacecraft in July 1975. This mission represented a fitting end to the space race.

American belief in the technical virtuosity of NASA, an agency that could accomplish any task assigned it, can be traced directly to the experience of Apollo and its legacy of success. The success in reaching the Moon established a popular conception that one could make virtually any demand and the space agency would deliver. This has remained a powerful image in American culture.

Despite tragedies along the way, including the near-disaster of *Apollo 13* and the *Challenger* and *Columbia* accidents that killed fourteen astronauts after the space race, most of the public remains convinced that NASA has the capability to succeed at whatever it attempts. The Moon landings established that image in the American mind, and it has resisted tarnish despite the space agency's very public failures great and small.

Of course, there has also been concern about an undefined sense of decline in so many parts of recent American society. Some have expressed a desire to recapture the confidence and technological virtuosity that helped define the 1960s but has flagged since. Farouk El-Baz bemoaned that "the Apollo spirit of innovation and can-do attitude did not last long." He concluded: "This is why I believe that my generation has failed the American people in one respect. We considered Apollo as an enormous challenge and a singular goal. To us, it was the end game. We knew that nothing like it ever happened in the past and behaved as if it would have no equal in the future."

The technology required to reach the Moon was certainly more complex than anything ever attempted before, but it was firmly understood at the time that the program began. NASA engineers reasoned, first, that they needed a truly powerful rocket with a larger payload capacity than any envisioned before. As a second priority, they recognized the need for a spacecraft that could preserve the life of fragile human beings for at least two weeks; this included both a vehicle akin to a small submarine that could operate in space and a second spacecraft in the form of a space suit that would allow the astronauts to perform tasks outside the larger vehicle. Third, they needed some type of landing craft that would be able to operate in a lunar environment far different from anything found on or near Earth. Finally, they needed to develop the technologies necessary for guidance and control, communication, and navigation to reach the Moon.

In every case—and this proved critical—planners at NASA understood the nature of the technical challenges before them in reaching for the Moon, so they were able to chart a reasonable and well-defined technological development course for overcoming them.

Project Apollo was a triumph of management in meeting enormously difficult systems engineering, technological, and organizational integration requirements. For the generation of Americans who grew up during the 1960s watching NASA astronauts fly into space, beginning with fifteen-minute suborbital trajectories and culminating with six landings on the Moon, Project Apollo signaled in a very public manner what the nation could do when it set its mind to it. Television coverage of real space adventures was long and intense, the stakes high, and the risks of life enormous. There were moments of both great danger and high anxiety.

Indeed, the Moon-landing program came to exemplify the best Americans could bring to any challenge and has been routinely deployed to support the nation's sense of greatness. Actor Carroll O'Connor, portraying the character of Archie Bunker, a bigoted working-class American whose perspectives were more common in our society than many observers were comfortable admitting, perhaps said it best in an episode of *All in the Family* in 1971. Archie represented well how most Americans embraced the success of the Apollo program; he observed to a visitor to his house that he had "a genuine facsimile of the Apollo 14 insignia. That's the thing that sets the US of A apart from . . . all them other losers." In very specific terms, Archie Bunker encapsulated for many what set the United States apart from other nations: success in spaceflight.

More recently, another reference from popular culture points up the lasting nature of this sense of success granted the nation through its Apollo Moon landings. In the critically acclaimed television situation comedy *Sports Night*, about a team that produces a nightly cable sports broadcast, an episode in 2001 included a telling discussion of space exploration. The fictional sports show's executive producer, Isaac Jaffee, played by African-American actor Robert Guillaume, is recovering from a stroke and disengaged from the daily hubbub of putting together the nightly show. His producer, played by Felicity Huffman, keeps interrupting him as he reads a magazine about space exploration. Isaac tells her, "They're talking about bio-engineering animals and terraforming Mars. When I started reporting Gemini missions, just watching a Titan rocket liftoff was a sight to see." Jaffee affirms his basic faith in NASA to carry out any task in space exploration. "You put an X anyplace in the solar system," he says, "and the engineers at NASA can land a spacecraft on it."

Each Apollo mission captured the essence of American technological prowess. If there is one hallmark of the American people, it is their enthusiasm for technology and what it can help them to accomplish. Historian Perry Miller wrote that the Puritans of New England used technology to transform a wilderness into their "City upon a Hill." They "flung themselves in the technological torrent," Miller wrote; "how they shouted with glee in the midst of the cataract, and cried to each other as they went headlong down the chute that here was their destiny." Since that time the United States has been known as a nation of technological system builders who could use this ability to create great machines of wonder, and the components of their operation. Perceptive foreigners might be enamored with American political and social developments, with democracy and pluralism, but they are more taken with U.S. technology. Not because it was intrinsically better, although that might be the case, but because it represented a communion of the human spirit with the eternal sublime of the universe. Apollo embodied that in ways rarely seen elsewhere. The technological virtuosity remains to this day. It has long supported an emphasis on national greatness and exceptionalism and offers solace in the face of other setbacks. At a basic level the Moon landings provided the impetus for the perception of NASA as a successful organization, and of the United States as the world leader in science and technology.

Scientific Return

"The direct scientific result of the Apollo Program, viewed collectively, can be summarized as fundamental new knowledge of the Moon, the Sun, and the Earth, and of the behavior of living and inanimate systems in the microgravity environment provided by orbiting spacecraft and space stations," wrote NASA scientist Paul D. Lowman in 1999. That was an understatement. From a scientific perspective Apollo opened the Moon to understanding as never before. The landforms of the Moon, systematically investigated by scientists with data from the lunar expeditions, advanced knowledge of the regolith, volcanism, tectonic actions, impacts, and the creation of the lunar face. In this context, scientists came to appreciate in a new way the origin of the Moon and the early history of the solar system.

In the end, executing the science of the Apollo program involved three hugely significant efforts carried out by the science community—landing site selection, instrument and experiment selection, and training of astronauts for scientific fieldwork. The first of these was arguably the most important. While the scientists had been planning since 1962 for investigations during each landing, they also got involved at the beginning with the Apollo Site Selection Board established at NASA headquarters in August 1965. This board served as the primary vehicle for determining where each mission to the Moon would land, and therefore the nature of scientific observation and experimentation permitted. It was always a contentious, but necessary, activity that successfully reached consensus on a range of geologically interesting landing sites.

The second major area that caused contention in the program was the definition of science experiments on the lunar surface. Ongoing debates about the size and mass of experiments, as well as their power requirements, roiled the mission planning efforts throughout the middle 1960s. The scientists agreed that the first investigations should relate to geology (especially sample collection), geochemistry, and geophysics. They also agreed that the early landings should focus on returning as many diverse lunar rock and soil samples as was feasible, deployment of long-lasting surface instruments, and geological exploration of the immediate landing areas by each crew. These could be expanded later to include surveys of the whole Moon and detailed studies of specific sites in the equatorial belt.

Finally, as we have seen, the scientific community helped prepare the Apollo astronauts for geological fieldwork on the Moon. Despite the relative rarity of academically trained geologists in the astronaut corps—recall that the only Moonwalker was the Harvard Ph.D. in geology Harrison Schmitt, and that other geologists such as Brian O'Leary resigned from the program in disgust—the scientists nonetheless made every effort to guide the flight crews, making certain that when they reached the lunar surface, they were ready to perform their assigned tasks. Their success was a triumph for the scientists as well as the astronauts. While some astronauts deplored this emphasis, others embraced the scientific task. In the end, between 1964 and the times of the various missions, the crews undertook classroom study and fieldwork in a variety of settings to prepare for their time on the lunar surface. In the end, most would agree that the flight crews had sufficient formal education to earn the equivalent of a master's degree in geology.

A key question answered by the Apollo science program involved the origin of the Moon. Before Apollo three principal theories existed about its origins:

1. Fission, a theory holding that the Moon had split off from Earth;
2. Co-accretion, a theory that proclaimed that the Moon and Earth formed at the same time from the Solar Nebula;
3. Capture, a theory that held that the Moon formed elsewhere and was subsequently drawn into orbit around Earth.

These theories all lost adherents because of the Apollo samples. After a decade of analysis, at an October 1984 conference of lunar scientists in Kona, Hawaii, a consensus emerged rather unexpectedly that the Moon had been formed by debris from a massive collision—the "big whack"—from a very large object (as big as Mars and named after the fact Theia) about 4.6 billion years ago. This "big whack" theory explained well what had been learned about the geology of the Moon during the Apollo program. While there are still details to be worked out, the impact theory is now widely accepted. Lunar scientists are eager to return to the Moon to answer additional questions about the Moon's origins.

Most scientists would probably agree with Ph.D. geologist and *Apollo 17* astronaut Harrison Schmitt, who recalled that we learned because of Apollo that "the Moon moves through space as an ancient text, related to the history of the Earth only through the interpretations of our minds, and, as the modern archive of our sun, recording in its soils much of immediate importance to man's future well-being."

The Space Race: Pride and Prestige

Central to any discussion of the space race is its role as an engine of national pride and international prestige for the United States in the context of Cold War rivalries. *Prestige*, for all its ubiquity in the literature of human spaceflight, is an imprecise term, which perhaps obscures more than it illuminates. It signifies a demonstration of national superiority over a rival. Both the Americans and Soviets pursued prestige. But this superiority has many facets and audiences. It both elicits a "gut-level" reaction and calls for a more sophisticated explication. It is driven by politics of many sorts—international, bureaucratic, and domestic—none of them sufficient on its own to explain the primacy of human spaceflight in American and Soviet cultures, but all complexly intertwined.

There may well be four distinct attributes of the issue of pride and prestige in the space race:

- Prestige on the international stage, using the space race as a means for influencing the attitudes of nonaligned populations toward either the Soviet Union or the United States;
- Pride at the national level, drawing the nation and its many peoples, priorities, and perspectives together;
- Definition of national identity, introducing important ingredients into a national narrative of exceptionalism;
- Valorization of the idea of progress, with the space race a symbol for national forward thinking.

This application of prestige is a classic application of what analysts often refer to as "soft power." Coined by Harvard University professor Joseph S. Nye, the term defined an alternative to threats and other forms of "hard power" in international relations. As Nye contends:

> Soft power is the ability to get what you want by attracting and persuading others to adopt your goals. It differs from hard power, the ability to use the carrots and sticks of economic and military might to make others follow your will. Both hard and soft power are important . . . but attraction is much cheaper than coercion, and an asset that needs to be nourished.

Such activities as the space race represented a form of soft power for both rivals, the ability to influence other nations through intangibles such as an impressive show of technological capability. It granted to the nation achieving successes an authenticity and gravitas not previously enjoyed among the world community. At sum, this was an argument buttressing the role of spaceflight as a means of enhancing prestige on the world stage.

There is no question but that all the human spaceflight efforts were initially about establishing U.S. or Soviet primacy in technology. Spaceflight served as a surrogate for war, with the United States and the Soviet Union in a head-on contest of technological virtuosity. The desire to win international support for the "American way" became the raison d'être for the Apollo program, and it served that purpose far better than anyone imagined when the program was first envisioned. Apollo became primarily a Cold War initiative that helped to demonstrate the mastery of the United States before the world. At the height of the Apollo Moon landings, world opinion had shifted overwhelmingly in favor of the United States. The importance of Apollo as an instrument of U.S. foreign policy—not necessarily identical with national prestige and geopolitics, but closely allied—should not be mislaid in this discussion. It served, and continues to serve, as an instrument for projecting the image of a positive, open, dynamic American society abroad.

The Space Race and the Idea of Progress

The iconic space race, especially the Moon landings, served very specific needs for both the United States and the Soviet Union, and it has largely been mobilized to bolster the stature of these nations in the period since. This represents a fulfillment of the dominant narrative of Soviet/Russian and American triumph, exceptionalism, and success. Through the process it served as an exemplar of a grand visionary concept for human exploration and progress. Given this observation, the space race has been celebrated as an investment in technology, science, and knowledge that would enable humanity to do more than just dip its toes in the cosmic ocean, to become a truly spacefaring people.

It is somewhat trite to suggest that America was founded on the idea of progress, and that progress remains both an amorphous concept and one central to American national identity. In the 1830s an astute French interpreter of United States society, Alexis de

Tocqueville, observed that Americans had a "lively faith in human perfectibility," and that as a society they believed they were "a body progressing" rather than one that either declined or remained stable.

If anything, Tocqueville understated this principle, for the concept of America as a Utopia in process has permeated the national ideology since before the birth of the Republic. From Puritan leader John Winthrop's "City upon a Hill," to Thomas Jefferson's stirring statement in the Declaration of Independence that people must work to ensure that all receive their unalienable rights of "Life, Liberty and the pursuit of Happiness," to the New Deal of Franklin D. Roosevelt in the 1930s and the Great Society of Lyndon B. Johnson in the 1960s, progress has been a major subtext of every aspect of American life.

The same was true of the Soviet Union. The Bolshevik Revolution of 1917 sought to create a utopian world in which all people contributed what they could to the welfare of all and received what they needed in return. In practice this did not work, but the idea of progress toward equanimity and equality was everywhere apparent.

Of course, the way these ideas have evolved over time has changed in relation to the larger society, and space exploration, especially the race to the Moon, evinced these cherished conceptions. As political scientist Taylor E. Dark III has argued:

> The idea of progress has typically advanced three claims: 1. There are no fundamental limits on the human capacity to grow, however growth is defined; 2. Advancements in science and technology foster improvements in the moral and political character of humanity; and, 3. There is an innate directionality in human society, rooted in societal, psychological, or biological mechanisms, that drives civilization toward advancement. American believers in progress quickly embraced space travel, viewing it as a vindication of the doctrine's original claims

> about the near-inevitability of human improvement. With space travel understood in this fashion, the fate of the space program took on a far greater meaning than developments in other areas of technological endeavor, as it became symbolic of the entire directionality of human civilization.

Although progress had been present earlier in the works of such space advocates as the Russian Konstantin E. Tsiolkovsky, the American Robert H. Goddard, and the German Wernher von Braun, after the conclusion of the space race in the early 1970s space enthusiasts believed they were on the verge of a new golden age in which anything could be accomplished. Apollo raised the hopes of those dreaming of great human progress in space. Its transcendental qualities were not lost on those who believed that humanity could eventually attain this end.

Movement into space, first with exploring expeditions and later with colonies, offered an opportunity for humanity to move outward and start anew on an untouched planet. The space race had shown it was possible. It suggested that the spacefaring nations had both the capability and the wherewithal to accomplish truly astounding goals. All it needed was the will. As Senator Abraham Ribicoff (D-CT) mused in 1969, "If men can visit the Moon—and now we know they can—then there is no limit to what else we can do. Perhaps that is the real meaning of Apollo 11."

We see this in the imagery of the space race as well. The essence of progress present in space race photography is unmistakable, along with the dominant narrative of national triumph, exceptionalism, and success so much a part of the interpretation of space exploration. From an advertising perspective, the linkage of humanity to this grand endeavor was an easy sell. Is it any wonder that it would be central to positive elements of human progress since the 1960s?

Apollo as Nostalgia

Decades later, the Apollo program represents a powerful incarnation of nostalgia for the excitement of seeing astronauts walk on the Moon. While there were setbacks, the experience of the Apollo years, 1961 to 1972, included more triumph than tragedy, more heroic sacrifice, more strenuous effort than many wars, and certainly much more daring than the years of space exploration since. Many remember Apollo as an effort wrought with high drama and excitement. More important, Apollo was a government program that succeeded. Twelve American astronauts did indeed land on the Moon and return safely to Earth, the first of them "before this decade is out," as John F. Kennedy had directed. All of this happened while the United States underwent social revolution, suffered defeat in Vietnam, and undertook the Great Society programs, many of which have since been widely branded as flawed, if not failed. And NASA accomplished Apollo within the confines of its overall budget targets. All this prompts many to reflect on the episode with nostalgia and longing for a return to a simpler time.

Apollo nostalgia manifests itself in several ways. It may be found in numerous popular conceptions of the program, especially in film, literature, music, theater, and advertising. In each of these arenas three great themes play out in evoking the past of Apollo. First, reaching for the Moon represented a spiritual quest, a purification of humanity, and a search for absolution and immortality. Because of this, much of the nostalgia for Apollo has all the trappings of a religion. Second, Apollo represented the next step in human evolution, and carried with it a Darwinian overtone of "survival of the fittest"; some look back on the winding down of the program as a missed opportunity. Third, and perhaps most important, Apollo nostalgia harks back to an era of the early 1960s in which order ruled and all seemed in its place. Clearly, this nostalgia is more a perception than a reality, but it bespeaks a belief about the period before the turmoil of the later 1960s and celebrates the place of Apollo coming out of an earlier time and circumstance. Whether appropriate or not—and in reality it was less so than many perceive—the dominance of the white male elite, the sense that Americans could accomplish anything they set their minds to, and the naïve emphasis that there existed a fundamental unity of American values and attitudes represented a powerful aphrodisiac for a lost past. Most important for reinforcement of this issue, the system worked and in memory enjoyed efficiencies lost in a postmodern, multicultural setting. This longing for a distant, dimly remembered, and ultimately inaccurate past represents perhaps the most troubling aspect of Apollo nostalgia.

Apollo as spiritual quest found expression from the unveiling of the program. At a uniquely oblique angle Project Apollo represented the incarnation of a new religious tradition. It evoked, in a metaphorical and absolutist sense, emotions of awe, devotion, omnipotence, and, most important, redemption for humanity. It embodied a new clerical caste (the engineers and especially the astronauts), arcane rituals (Mission Control and other operational activities) that were deeply mythical as well as possessing a higher purpose, a language of devotion (NASA jargon invoked by both practitioners and acolytes/enthusiasts), articles of faith, and a theology of salvation that allowed humanity to reach beyond Earth and settle the cosmos. The promise of a utopian Zion on a new world, coupled with immortality for the species, resonates through every fiber of the space-exploration community. For those who embraced this idea of a new Zion, as has always been the case for adherents throughout history, the present culture was unrighteous and unnerving. They chose to escape, and leaving the planet could be the ultimate form of escape. Wernher von Braun, one example among many, viewed Apollo as a millenarian new beginning for humankind. These deep-seated convictions energized space exploration and the subjugation of the universe from before the dawn of the space age.

But while many Apollo advocates presented the program in explicitly spiritual terms, countless others used secular language to express religious ideals. American novelist Ray Bradbury once commented in a fashion reminiscent of a jeremiad: "Too many of us have lost the passion and emotion of the remarkable things we've done in space. Let us not tear up the future, but rather again heed the creative metaphors that render space travel a religious experience. When the blast of a rocket launch slams you against the wall and all the rust is shaken off your body, you will hear the great shout of the universe and the joyful crying of people who have been changed by what they've seen." Bradbury firmly believed that no one leaves a space launch untransformed. Like the Eucharist, the ritual of the launch offers a recommitment to the endeavor and a symbolic cleansing of the communicant's sole. The experience, as he commented repeatedly, is both thrilling and sanctifying.

Norman Mailer's critique of Apollo, *Of a Fire on the Moon* (1969), eloquently observed that these missions were really about seeking to become one with God: "They don't know what to do when they get there. The fact that it's technological is what's wrong with it. It's too exclusively technological. People are sick to death of technology. The technologists themselves are wondering how they can control technology before technology wipes out the Earth. So what we're looking for at this time in human history is an enlargement of human consciousness, a rediscovery of spiritual values to which we can adhere because they deepen us."

That same spiritual quest, coupled with the technological enterprise, is present in the 1995 feature film *Apollo 13*. That mission, in 1970, when an explosion crippled a lunar-landing mission and NASA nearly lost astronauts Jim Lovell, Fred Haise, and Jack Swigert, has been recast as one of NASA's finest hours, a successful failure. Fifty-six hours into the flight an oxygen tank in the Apollo service module ruptured and damaged several of the power, electrical, and life-support systems. People around the world watched and waited and hoped and prayed as NASA personnel on the ground and the crew worked to find a way safely home. It was a close call, but the crew returned safely on April 17, 1970. The near-disaster at the same time solidified in the popular mind NASA's collective genius and prompted reconsideration of the propriety of the whole effort. While one must give the NASA flight team high marks for perseverance, dedication, and an unshakable belief that they could bring the crew home safely, it is quite strange that no one seems to realize that the mission had already failed, and failed catastrophically, by the time of the accident. The fact that *Apollo 13* is now viewed as one of NASA's shining moments says much about the ability of humanity to recast historical events into meaningful morality plays.

From this skewed perspective, *Apollo 13* became a vehicle for criticism of the social order that emerged from the 1960s and a celebration of an earlier age. When the film appeared in 1995, reviewer John Powers, writing for the *Washington Post*, commented on its incessant nostalgia for "the paradisiacal America invoked by Ronald Reagan and Pat Buchanan—an America where men were men, women were subservient, and people of color kept out of the way." In addition, Powers wrote, "Its story line could be a Republican parable about 1995 America: A marvelous vessel loses its power and speeds toward extinction, until it's saved by a team of heroic white men." If anything, Powers underemphasized the white America evoked in *Apollo 13*. Only two women have speaking parts of substance. Kathleen Quinlan as Marilyn Lovell, wife of the *Apollo 13* commander, steadfastly offers proud support while privately fearing the worst. Mary Kate Schellhardt as the Lovells' daughter functions as a spokesperson for the least important elements of the social revolution, complaining at one point in a shrewish voice that the Beatles have broken up and her world has accordingly collapsed.

The heroes of *Apollo 13* are the geeks of Mission Control, while the astronauts aboard the spacecraft are spirited but essentially emasculated characters to be saved. Lovell, Haise, and Swigert must wait to be rescued in a manner not unlike Rapunzel, as an active helper but unable to accomplish the task alone. As historian Tom D. Crouch wrote of this film's depiction of the "studs" in Mission Control:

> The real heroes of this film are either bald or sporting brush cuts; wear thick glasses; are partial to rumpled short sleeve shirts; and chain-smoke an endless string of cigarettes, cigars, and pipes. For all of that, these slide rule–wielding technonerds solve all of the difficult problems required to bring the crew home. They are, in the words of one of the astronauts portrayed in the film, "steely-eyed missile men."

Apollo 13 the film, accordingly, venerates a long-past era in American history. Indeed, it may have been an era already gone by the time of the actual mission in 1970. It is a hallowing of masculinity in a nostalgic context.

A recent study completed for NASA concluded that representation of space exploration in films is highly nostalgic, and Apollo fuels that perception:

> As a group, the public entertainments we tend to buy into are either nostalgic visions of the "space race" period ("The Right Stuff," "Apollo 13," "From the Earth to the Moon") or fantasies reflecting the romantic imagination of the Flash Gordon/Buck Rogers era ("Star Wars" rather than "Star Trek"). These are the visions people support in the most meaningful way possible: with their time and dollars. . . . Boomers have a great nostalgic affection for NASA, but their own priorities have

> shifted from a future focus to maintaining what they have. They see money spent on space exploration as threatening their future entitlements.

At a sublime level, Apollo nostalgia may serve as a trope for a larger lack of interest in the future expressed by Americans in the first part of the twenty-first century. A shifting cultural center of gravity toward maintenance of the status quo has become common.

Conclusion

The space race in general, the Moon landings especially, should be viewed as a watershed in world history. It was an endeavor that demonstrated both the technological and economic brilliance of the nations involved and established scientific-technological mastery as the fundamental demarcation of a successful nation-state. It had been an enormous undertaking, costing more than $50 billion (about $250 billion in 2018 dollars), with only the building of the Panama Canal rivaling the space race in size as the largest nonmilitary technological endeavor ever undertaken. Several important legacies of the space race should be remembered. First, and probably most important, the Americans won the race despite an early and persistent lead by the Soviets through the first half of the 1960s.

In both nations, but especially in the United States, the space race successfully accomplished the political goals for which it had been created. Kennedy and Khrushchev had been dealing with one Cold War crisis after another by 1961, when the Americans announced the Moon race. Upon its successful conclusion, at the time of the

Apollo 11 landing, Mission Control in Houston flashed the words of President John F. Kennedy announcing the Apollo commitment on its big screen. Those phrases were followed with these: "TASK ACCOMPLISHED, July 1969." No greater understatement could probably have been made. Any assessment of Apollo that does not recognize the accomplishment of landing Americans on the Moon and safely bringing them home before the end of the 1960s is incomplete and inaccurate, for that was the primary goal of the undertaking.

Second, the space race demonstrated the triumph of management in meeting enormously difficult systems engineering, technological, and organizational integration requirements. Sergei Korolev effectively combated internal bureaucratic infighting to keep the Soviet program on track until his death in January 1966. No one else possessed the stature and gravitas to keep those forces in check thereafter. Korolev's longtime lieutenant, Vasily Mishin, proved no match for the politically connected Valentin Glushko, and the Soviet program declined into disorder. Its successes in the late 1960s were testament to the technical competence of the engineers and scientists working in the program. It raises the question, might there have been a different outcome had not Korolev died so suddenly? It is easy to imagine that he might have rallied the resources and lined up the support in the Kremlin for a push to beat the Americans to the Moon.

As it turned out, that organizational flair characterized NASA and its senior leadership. The NASA administrator at the height of the program between 1961 and 1968, James E. Webb, kept his team on track, provided his scientists and engineers with the resources they required, and ensured political support where necessary. NASA proved that the space race was much more a management exercise than anything else, and that the technological challenge, while sophisticated and impressive, was within grasp. More difficult was ensuring that those technological skills were properly managed and used. Webb's contention was confirmed by Apollo.

Third, the space race forced the people of the world to view the planet Earth in a new way. It helped to galvanize the modern environmental movement and give legs to the quest for an understanding of the geophysical features of our planet. Perhaps just as important, NASA pursued critical Earth-science objectives in the 1960s. Indeed, at a fundamental level NASA—along with the National Oceanic and Atmospheric Administration—became the critical component in the 1960s of the origins of a new scientific discipline that emerged in the United States, Earth system science. During that decade NASA developed the critical technology that made possible the convergence of many different scientific disciplines into a single system of investigation. It presided over the rise of interdisciplinarity in the various sciences focusing on understanding the Earth. As such, it incorporated understandings of how the atmosphere, ocean, land, and biospheric components of Earth interacted as an integrated system. This resulted from studies of the interaction between the physical climate system and biogeochemical cycles.

Very early the role of humans in this process emerged as NASA pursued research with its Landsat satellites to demonstrate changes in land use and ground cover. Only the analysis of data obtained through both *in situ* observations and from remotely sensed observations, as well as the development of sophisticated ocean-atmosphere-land models, made this possible. Not until the space race did a fundamental ingredient of this process emerge in the use of satellites. By the 1970s Earth system science offered a foundation for understanding and forecasting changes in the global environment and regional implications.

Finally, the Apollo program, although an enormous achievement, left a divided legacy for both the American and Soviet efforts at space exploration. The space race has been perceived as a golden age that would be forever recognized as the ultimate achievement of humanity, but as it recedes into the past, it looks more like an anomaly than the norm.

Apollo, for example, has taken on mythical characteristics, but ones that are bittersweet. Historian Alex Roland captured this Apollo myth best: as a retelling of U.S. exceptionalism for a specific purpose. In this setting, it is not so much about history as it is the communication of "tribal rituals, meant to comfort the old and indoctrinate the young." He noted in 1994:

> All the exhilarating stories are here: the brave, visionary young President who set America on a course to the moon and immortality; the 400,000 workers across the nation who built the Apollo spacecraft; the swashbuckling astronauts who exuded the right stuff; the preliminary flights of Mercury and Gemini —from Alan Shepard's suborbital arc into space, through John Glenn's first tentative orbits, through the rendezvous and spacewalks of Gemini that rehearsed the techniques necessary for Apollo. There is the 1967 fire that killed three astronauts and charred ineradicably the Apollo record and the Apollo memory; the circumlunar flight of Christmas 1968 that introduced the world to Earth-rise over the lunar landscape; the climax of Apollo 11 and Neil Armstrong's heroic piloting and modest words, "that's one small step for a man, one giant leap for mankind"; the even greater drama of Apollo 13, rocked by an explosion on the way to the moon and converted to a lifeboat that returned its crew safely to Earth thanks to the true heroics of the engineers in Houston; and, finally, the anticlimax of the last Apollo missions.

Roland finds an epic aura of Apollo in this recitation of the voyages of discovery. The missions, however, turned into a dead end rather than a new beginning, and no amount of heroic prose could overcome that unforeseen plot twist. The Apollo project was, therefore, an anomaly in the national decision-making process. The perception of a "golden age" of space exploration has been difficult to overcome.

Something that neither the United States nor the USSR understood as the space race was under way is that such an effort would never be repeated. Racing to the Moon was an anomaly in human history. It had more in common with the building of the pyramids of Egypt and the great cathedrals of Europe than of anything that might be considered normal public policy. The symbolism of the space race has held special appeal for the true believers of space exploration. To them, it suggested that space exploration deserved special consideration. The decision to engage in expansive space-exploration activities was sui generis, and not to be questioned. Tragically, this illusion has held sway for more than a generation since the last landing on the Moon. There is no second space race in the foreseeable future.

For Further Review

Books

Aldrin, Buzz. *Men from Earth*. New York: Bantam, 1989. Provides an intimate account of how NASA accomplished the national goal of putting an American on the Moon before the end of the decade. Interweaves the story of the U.S.-Soviet race to reach the Moon with the author's firsthand experience flying both the Gemini and Apollo missions during the height of the space race. Aldrin's recounting of his two spaceflights is compelling, especially the account of the nearly aborted *Apollo 11* lunar landing.

———. *Return to Earth*. New York: Bantam, 1973. Describes the celebrity associated with being the second human on the Moon and the author's struggles with alcoholism and depression in the early 1970s. Aldrin writes about the pressure to keep the stress and day-to-day problems inside, and its effect on his marriage, which ended in divorce.

Allday, Jonathan. *Apollo in Perspective: Spaceflight Then and Now*. New York: Institute of Physics Publications, 1999. Takes a retrospective look at the Apollo space program and the technology that was used to land an American on the Moon as a means to explain the basic physics and technology of spaceflight. Also conveys the huge technological strides that were made and the dedication of the people working on the program.

Armstrong, Neil A., et al. *First on the Moon: A Voyage with Neil Armstrong, Michael*

Collins, and Edwin E. Aldrin, Jr. Written with Gene Farmer and Dora Jane Hamblin. Epilogue by Arthur C. Clarke. Boston: Little, Brown, 1970. This is the "official" memoir of the *Apollo 11* landing mission to the Moon in 1969. It was prepared by the ghostwriters Farmer and Hamblin from information made available exclusively to them through a somewhat infamous Time-Life/Field Enterprises contract that excluded the rest of the media from contact with the astronauts' families. Contains much personal information about the astronauts that is not available elsewhere.

Bean, Alan L. *Apollo: An Eyewitness Account by Astronaut/Explorer Artist/Moonwalker.* Shelton, CT: Greenwich Workshop, 1998. This is a large-format discussion of Apollo written by an *Apollo 12* crewmember and illustrated with his unique artwork.

Beattie, Donald A. *Taking Science to the Moon: Lunar Experiments and the Apollo Program.* Baltimore: Johns Hopkins University Press, 2001. A history of the lunar science undertaken during Project Apollo. Beattie gives a firsthand account of efforts by NASA scientists to do more to include science payloads on Apollo missions despite opposition from mission engineers, who envisioned a direct round-trip shot with as much margin for error as possible.

Benjamin, Marina. *Rocket Dreams: How the Space Age Shaped Our Vision of a World Beyond.* New York: Free Press, 2003. The author ruminates on the scarred spaceflight culture that Apollo created and the later space program destroyed. She visits Roswell, New Mexico, with its alien kitsch, and the Kennedy Space Center in Cocoa Beach, Florida, with its gigantic rocket assembly buildings and launch complexes and reminders of the heyday of Apollo, when humans went to the Moon.

Benson, Charles D., and William Barnaby Faherty. *Moonport: A History of Apollo Launch Facilities and Operations.* Washington, DC: NASA SP-4204, 1978. An excellent official history of the design and construction of the lunar launch facilities at Kennedy Space Center. This book was reprinted in 2001 as a two-volume paperback by University Press of Florida under the titles *Gateway to the Moon: Building the Kennedy Space Center Launch Complex* and *Moon Launch! A History of the Saturn-Apollo Launch Operations.*

Bilstein, Roger E. *Stages to Saturn: A Technological History of the Apollo/Saturn Launch Vehicles.* Washington, DC: NASA SP-4206, 1980, rpt. ed. 1996. This thorough and well-written book gives a detailed but highly readable account of the enormously complex process whereby NASA and especially the Marshall Space Flight Center under the direction of Wernher von Braun developed the launch vehicles used in the Apollo program ultimately to send twelve humans to the Moon. Based on exhaustive research and equipped with extensive bibliographic references, this book comes as close to being a definitive history of the Saturn rocket program as is likely ever to appear. Reprinted in 2002 by University Press of Florida.

Borman, Frank, with Robert J. Serling. *Countdown: An Autobiography*. New York: William Morrow, Silver Arrow, 1988. Written to appear on the twentieth anniversary of the first lunar landing, this autobiography spans much more than the Apollo program. It recounts Borman's life in aeronautics, first as a military flier, then as a test pilot, and finally as president of Eastern Airlines.

Brooks, Courtney G., James M. Grimwood, and Loyd S. Swenson, Jr. *Chariots for Apollo: A History of Manned Lunar Spacecraft*. Washington, DC: NASA SP-4205, 1979. Based on exhaustive documentary and secondary research as well as 341 interviews, this well-written volume covers the design, development, testing, evaluation, and operational use of the Apollo spacecraft through July 1969.

Burgess, Colin, ed. *Footprints in the Dust: The Epic Voyages of Apollo, 1969–1975*. Lincoln: University of Nebraska Press, 2010. This book covers the flights of the Apollo program from *Apollo 11* through the Apollo-Soyuz mission in 1975.

Burgess, Colin, and Kate Doolan, with Bert Viz. *Fallen Astronauts: Heroes Who Died Reaching for the Moon*. Lincoln: University of Nebraska Press, 2017. This book tells the stories of the astronauts who died while employed by NASA.

Burrows, William E. *This New Ocean: The Story of the First Space Age*. New York: Random House, 1998. A comprehensive history of spaceflight, which tries to do too much but succeeds in explaining the political, technical, scientific, economic, and cultural history of humanity's recent adventure in space.

Cadbury, Deborah. *Space Race: The Epic Battle between America and the Soviet Union for the Dominion of Space*. New York: Harper Perennial, 2007. A journalistic account of the race to the Moon.

CBS News. *10:56:20 PM EDT, 7/20/69: The Historic Conquest of the Moon as Reported to the American People*. New York: Columbia Broadcasting System, 1970. As the title suggests, this is an attempt to capture in print and pictures the reporting on humankind's first landing on the Moon during *Apollo 11*. More useful in capturing the immediacy of the moment than in providing an historical assessment of the event and its significance.

Cernan, Eugene, with Donald A. Davis. *The Last Man on the Moon: Astronaut Eugene Cernan and America's Race in Space*. New York: St. Martin's, 1999. Gene Cernan, the last person to walk on the Moon, presents this memoir that starts with his childhood days outside Chicago, through college life at Purdue and his early career as a naval aviator, culminating with his career as an astronaut with his flight to the Moon on *Apollo 17*.

Chaikin, Andrew. *A Man on the Moon: The Voyages of the Apollo Astronauts*. New York: Viking, 1994. One of the best books on Apollo, this work emphasizes the exploration of the Moon by the astronauts between 1968 and 1972.

Chapman, Richard L. *Project Management in NASA: The System and the Men*. Washington, DC: NASA SP-324, 1973. Based on almost 150 interviews and contributions by NASA officials, this volume provides a useful look at NASA's project management system that contributed significantly to the success of the Apollo program.

Collins, Michael. *Carrying the Fire: An Astronaut's Journeys*. New York: Farrar, Straus and Giroux, 1974. This is the first candid book about life as an astronaut, written by the member of the *Apollo 11* crew who remained in orbit around the Moon. The author comments on other astronauts, describes the seemingly endless preparations for flights to the Moon, and assesses the results.

Compton, W. David, and Charles D. Benson. *Living and Working in Space: A History of Skylab*. Washington, DC: NASA SP-4208, 1983. The official NASA history of Skylab, an orbital workshop placed in orbit in the early 1970s.

———. *Where No Man Has Gone Before: A History of Apollo Lunar Exploration Missions*. Washington, DC: NASA SP-4214, 1989. This clearly written account traces the scientific aspects of the Apollo program.

Cortright, Edgar M., ed. *Apollo Expeditions to the Moon*. Washington, DC: NASA SP-350, 1975. This large-format volume, with numerous illustrations in both color and black and white, contains essays by numerous luminaries ranging from NASA Administrator James E. Webb to astronauts Michael Collins and Buzz Aldrin.

Ezell, Edward Clinton, and Linda Neuman Ezell. *The Partnership: A History of the Apollo-Soyuz Test Project*. Washington, DC: NASA SP-4209, 1978. An outstanding detailed study of the effort by the United States and the Soviet Union in the mid-1970s to conduct a joint human spaceflight.

Fowler, Eugene. *One Small Step: Project Apollo and the Legacy of the Space Age*. New York: Smithmark, 1999. This is a large-format "coffee table" history. Rather than focus just on the Apollo program itself, the book splits its contents almost evenly between the history of Apollo and the cultural impact of the space age.

Fries, Sylvia D. *NASA Engineers and the Age of Apollo*. Washington, DC: NASA SP-4104, 1992. This book is a sociocultural analysis of a selection of engineers at NASA who worked on Project Apollo. The author makes extensive use of oral history, providing both a significant appraisal of NASA during its "golden age" and important documentary material for future explorations.

Goldstein, Stanley H. *Reaching for the Stars: The Story of Astronaut Training and the Lunar Landing*. New York: Praeger, 1987. This is a detailed account of the development and management of the astronaut training program for Project Apollo.

Gray, Mike. *Angle of Attack: Harrison Storms and the Race to the Moon*. New York: Norton, 1992. This is a lively journalistic account of the career of Harrison Storms, president of the Aerospace Division of North American Aviation, which built the Apollo capsule. Because of the Apollo 204 fire that killed three astronauts in January 1967, Storms and North American Aviation got sucked into a controversy over accountability and responsibility. In the aftermath Storms was removed from oversight of the project. The most important aspect of this book is its discussion of the Apollo fire and responsibility for it from the perspective of industry. It lays the blame at NASA's feet and argues that Storms and North American were mere scapegoats.

Hall, Eldon C. *Journey to the Moon: The History of the Apollo Guidance Computer*. Reston, VA: American Institute of Aeronautics and Astronautics, 1996. A detailed history of the development of the pioneering guidance computer built for the Apollo lunar module by MIT's Draper Laboratory. The author was a senior participant in this effort.

Hallion, Richard P., and Tom D. Crouch, eds. *Apollo: Ten Years since Tranquility Base*. Washington, DC: Smithsonian Institution Press, 1979. This is a collection of essays developed for the National Air and Space Museum, commemorating the tenth anniversary of the first landing on the Moon, July 20, 1969. It consists of sixteen essays mostly written directly for the National Air and Space Museum by a variety of experts.

Hansen, James R. *First Man: The Life of Neil A. Armstrong*. New York: Simon and Schuster, 2005. This is the standard biography of Armstrong.

Hardesty, Von, and Gene Eisman. *Epic Rivalry: The Inside Story of the Soviet and American Space Race*. Washington, DC: National Geographic, 2007. A solid attempt to tell the story of the space race, written at the fiftieth anniversary of *Sputnik*.

Harford, James J. *Korolev: How One Man Masterminded the Soviet Drive to Beat America to the Moon*. New York: John Wiley, 1997. The first English-language biography of the Soviet "chief designer," who directed the projects that were so successful in the late 1950s and early 1960s in energizing the Cold War rivalry for space supremacy.

Harland, David M. *Exploring the Moon: The Apollo Expeditions*. Chichester, England:

Wiley-Praxis, 1999. This work focuses on the exploration and science missions carried out by Apollo astronauts while on the lunar surface.

Johnson, Stephen B. *The Secret of Apollo: Systems Management in American and European Space Programs*. Baltimore: Johns Hopkins University Press, 2002. This book skilfully interweaves technical details and fascinating personalities to tell the history of systems management in the United States and Europe. It is a very important work that uses Apollo as a key example.

Kauffman, James L. *Selling Outer Space: Kennedy, the Media, and Funding for Project Apollo, 1961–1963*. Tuscaloosa: University of Alabama Press, 1994. A straightforward history, but one that is quite helpful, of the public image-building efforts of NASA and the relation of that image to public policy.

Kelly, Thomas J. *Moon Lander: How We Developed the Lunar Module*. Washington, DC: Smithsonian Institution Press, 2001. An outstanding memoir of the building of the lunar module, written by the Grumman engineer who led the effort.

Kluger, Jeffrey. *Apollo 8: The Thrilling Story of the First Mission to the Moon*. New York: Henry Holt, 2017. A retelling of the *Apollo 8* mission through the eyes of the crew of the mission.

Kraft, Christopher C., with James L. Schefter. *Flight: My Life in Mission Control*. New York: E. P. Dutton, 2001. Full of anecdotes, this memoir of Mission Control in Houston is most entertaining.

Kranz, Gene. *Failure Is Not an Option: Mission Control from Mercury to Apollo 13 and Beyond*. New York: Simon and Schuster, 2000. A good memoir of Mission Control.

Lambright, W. Henry. *Powering Apollo: James E. Webb of NASA*. Baltimore: Johns Hopkins University Press, 1995. This is an excellent biography of James E. Webb, who served as NASA administrator between 1961 and 1968, the critical period in which Project Apollo was under way. During his tenure NASA developed the modern techniques necessary to coordinate and direct the most complex technological enterprise in human history, the sending of human beings to the Moon and bringing them safely back to Earth.

Launius, Roger D. *Apollo: A Retrospective Analysis*. Washington, DC: NASA SP-2004-4503, 1994, 2nd ed. 2004. A short study of Apollo's history with key documents.

———. *NASA: A History of the U.S. Civil Space Program*. Melbourne, FL: Krieger, 1994, rev. ed. 2001. A short book in the Anvil Series, this history of U.S. civilian space efforts consists half of narrative and half of documents. It contains three chapters on the Apollo program, but while coverage consists more of overview than detailed analysis, the approach is broadly analytical and provides the most recent general treatment of its topic.

Levine, Arnold S. *Managing NASA in the Apollo Era*. Washington, DC: NASA SP-

4102, 1982. A narrative account of NASA from its origins through 1969, this book analyzes key administrative decisions, contracting, personnel, the budgetary process, headquarters organization, relations with the Department of Defense, and long-range planning.

Liebergot, Sy, and David M. Harland. *Apollo EECOM: The Journey of a Lifetime*. Burlington, ON: Apogee, 2003. The autobiography of one of the key members of Mission Control in Houston during the Apollo program.

Light, Michael. *Full Moon*. New York: Alfred A. Knopf, 1999. In this book Michael Light has woven 129 of these stunningly clear images into a single composite voyage, a narrative of breathtaking immediacy and authenticity.

Lindsay, Hamish. *Tracking Apollo to the Moon*. New York: Springer Verlag, 2001. A history of the Apollo program from the perspective of an Australian involved in the tracking of the spacecraft that went to the Moon.

Logsdon, John M., gen. ed. *Exploring the Unknown: Selected Documents in the History of the U.S. Civil Space Program*. 6 vols. Washington, DC: NASA Special Publication-4407, 1995–2004. An essential reference work, these volumes print more than 700 key documents in space policy and its development throughout the twentieth century.

———. *John F. Kennedy and the Race to the Moon*. New York: Palgrave Macmillan, 2010. This study, based on extensive research in primary documents and archival interviews with key members of the Kennedy administration, is the definitive examination of John Kennedy's role in sending Americans to the Moon. Among other revelations, the author finds that after the Cuban missile crisis in 1962, JFK pursued an effort to turn Apollo into a cooperative program with the Soviet Union.

Lovell, Jim, and Jeffrey Kluger. *Lost Moon: The Perilous Voyage of Apollo 13*. Boston: Houghton Mifflin, 1994. After the 1995 film *Apollo 13*, no astronaut had more fame than Jim Lovell, commander of the ill-fated mission to the Moon in 1970. This book is his recollection of the mission and the record on which the theatrical release was based.

Mackenzie, Dana, *The Big Splat, or How Our Moon Came to Be*. Hoboken, NJ: John Wiley, 2003. A fine discussion of how the science of Apollo led to a new interpretation of the origins of the Moon.

Maher, Neil M. *Apollo in the Age of Aquarius*. Cambridge: Harvard University Press, 2017. A major reinterpretation of the Apollo program and its relationship to the counterculture of the 1960s.

Mailer, Norman. *Of a Fire on the Moon*. Boston: Little, Brown, 1970. One of the foremost contemporary American writers, Mailer was commissioned to write about the first lunar landing. The book reflects Mailer's 1960s countercultural mindset in meeting its antithesis, a NASA steeped in middle-class values and reverence for the American flag and culture.

Makemson, Harlen. *Media, NASA, and America's Quest for the Moon*. New York: Peter Lang, 2009. A study of media's reporting on the lunar program.

McCurdy, Howard E. *Inside NASA: High-Technology and Organization Change in the U.S. Space Program*. Baltimore: Johns Hopkins University Press, 1993. A major study showing change to the organizational culture from the Apollo era to the present.

———. *Space and the American Imagination*. Washington, DC: Smithsonian Institution Press, 1997. A pathbreaking study of the relationship between space and American culture.

McDougall, Walter A. . . . *the Heavens and the Earth: A Political History of the Space Age*. New York: Basic, 1985. This Pulitzer Prize–winning book analyzes the race to the Moon in the 1960s. The author, then teaching at the University of California, Berkeley, argues that Apollo prompted the space program to stress engineering over science, competition over cooperation, civilian over military management, and international prestige over practical applications. While he recognizes Apollo as a "magnificent achievement," he concludes that it was also enormously costly. Emphasizing the effect of space upon American society, this history focuses on the role of the state as a promoter of technological progress.

Mindell, David. *Digital Apollo: Human and Machine in Spaceflight*. Cambridge: MIT Press, 2011. An important study of the development of the Apollo guidance computer.

Mitchell, Edgar D., with Dwight Williams. *The Way of the Explorer: An Apollo Astronaut's Journey through the Material and Mystical Worlds*. New York: G. P. Putnam, 1996. A member of the *Apollo 14* crew, Mitchell presents a smooth blend of autobiography and exegesis, commenting at length on the experiments in extrasensory perception he conducted on the flight and on his spiritual journey since returning to Earth.

Mitchell, Edgar, and Ellen Mahoney, *Earthrise: My Adventures as an Apollo 14 Astronaut*. Chicago: Chicago Review Press, 2014. This is the inspiring and fascinating biography of the sixth man to walk on the Moon. Of the nearly seven billion people who live on Earth, only twelve have walked on the Moon, and Edgar Mitchell was one of them. *Earthrise* is a vibrant memoir for young adults featuring the life story of this internationally known *Apollo 14* astronaut. The book focuses on Mitchell's amazing journey to the Moon in 1971 and highlights the many steps he took to get there. In engaging and suspenseful prose, he details his historic flight to the Moon, describing everything from the very practical (eating, sleeping, and going to the bathroom in space) to the metaphysical (experiencing a life-changing connectedness to the universe).

Monchaux, Nicholas de. *Spacesuit: Fashioning Apollo*. Cambridge: MIT Press, 2011. This scintillating and innovative book explores layers of the space suit to tell the human story of its construction and use, as well as the stories of those who made and used it.

Montgomery, Scott L. *The Moon and the Western Imagination*. Tucson: University of Arizona Press, 1999. The author has produced a richly detailed analysis of how the Moon has been visualized in Western culture through the ages, revealing the faces it has presented to philosophers, writers, artists, and scientists for nearly three millennia.

Murray, Charles A., and Catherine Bly Cox. *Apollo: The Race to the Moon*. New York: Simon and Schuster, 1989. Rpt. ed., Burkittsville, MD: South Mountain, 2004. Perhaps the best general account of the lunar program, this history uses interviews and documents to reconstruct the stories of the people who participated in Apollo.

Neufeld, Michael J. *Von Braun: Dreamer of Space, Engineer of War*. New York: Alfred A. Knopf, 2007. This is the standard work on the life of the rocket pioneer and the godfather of the Saturn V rocket that took astronauts to the Moon.

Oberg, James E. *Red Star in Orbit*. New York: Random House, 1981. Written by one of the premier Soviet space watchers, this history of the Soviet space program is among the best published in English before the fall of the Soviet Union in 1989. Based mostly on Western sources, it describes what was then known of the Soviet Union's efforts to land a cosmonaut on the Moon before the U.S. Apollo landing in 1969.

Oliver, Kendrick. *To Touch the Face of God: The Sacred, the Profane, and the American Space Program, 1957–1975*. Baltimore: Johns Hopkins University Press, 2012. This is an underappreciated aspect of the ideology of human spaceflight. While historians have expended great effort to understand the influence of the Cold War in explaining the United States' embarkation in the difficult task of exploring space with humans, we have done little more than tangentially recognize that there seems to be something more to the support for human spaceflight than just practicality and realpolitik.

Oreskes, Naomi, and John Krige, eds. *Science and Technology in the Global Cold War*. Cambridge: MIT Press, 2014. An important collection of essays, especially Asif A. Siddiqi's "Fighting Each Other: The N-1, Soviet Big Science, and the Cold War at Home."

Orloff, Richard G., compiler. *Apollo by the Numbers: A Statistical Reference*. Washington, DC: NASA SP-2000-4029, 2000. An excellent statistical reference.

Paul, Richard, and Steven Moss. *We Could Not Fail: The First African Americans in the Space Program*. Austin: University of Texas Press, 2015. A major reinterpretation of the place of African-American engineers and scientists in the Apollo program.

Pellegrino, Charles R., and Joshua Stoff. *Chariots for Apollo: The Making of the Lunar Module*. New York: Atheneum, 1985. A popular though not always accurate discussion of the development of the lunar module by the Grumman Aerospace Corporation.

Poole, Robert. *Earthrise: How Man First Saw the Earth*. New Haven: Yale University Press, 2008. A pathbreaking book on the *Apollo 8* mission and the "Earthrise" photograph that captured the global imagination.

Reynolds, David West. *Apollo: The Epic Journey to the Moon, 1963–1972*. New York: Zenith, 2013, rpt. Featuring a wealth of rare photographs, artwork, and cutaway illustrations, the book recaptures the excitement of the United States' journey to the Moon.

Schirra, Wally, and Richard N. Billings. *Schirra's Space*. Annapolis: Naval Institute Press, 1995. Wally Schirra was the only one of the original seven NASA astronauts to command a spacecraft in all three pioneering space programs—Mercury, Gemini, and Apollo.

Scott, David Meerman, and Richard Jurek. *Marketing the Moon: The Selling of the Apollo Lunar Program*. Cambridge: MIT Press, 2014. An illustrated work on the sophisticated efforts by NASA and its many contractors to market the facts about space travel—through press releases, bylined articles, lavishly detailed background materials, and fully produced radio and television features—rather than push an agenda.

Shayler, David J. *Apollo: The Lost and Forgotten Missions*. Chichester, England: Springer-Praxis, 2002. A discussion of planning for the aborted *Apollo 18, 19,* and *20* missions.

Shepard, Alan, and Deke Slayton. *Moonshot: The Inside Story of America's Race to the Moon*. New York: Turner, 1994. Based on the recollections of two of the original Mercury Seven astronauts chosen in 1959, this book is a disappointing general history of human space exploration by NASA from the first flight in 1961 through the last Apollo landing in 1972.

Siddiqi, Asif A. *Challenge to Apollo: The Soviet Union and the Space Race, 1945–1974*. Washington, DC: NASA SP-2000-4408, 2000. The Soviet side of the race to the Moon. Reprinted as a two-volume paperback by University Press of Florida in 2003.

———. *The Red Rockets' Glare: Spaceflight and the Soviet Imagination, 1857–1957*. New York: Cambridge University Press, 2010. A seminal study of the origins of the Soviet space program.

Slayton, Donald K., and Michael Cassutt. *Deke! U.S. Manned Space, From Mercury to the Shuttle*. New York: Forge, 1995. This is the autobiography of one of the original Mercury Seven astronauts, selected in April 1959 to fly in space. Deke Slayton served as a NASA astronaut during Projects Mercury, Gemini, Apollo, Skylab, and the Apollo-Soyuz Test Project (ASTP), and while he was originally scheduled to pilot the Mercury-Atlas 7 mission, he was relieved of this assignment due to a mild occasional irregular heart palpitation discovered in August 1959. His only spaceflight took place in July 1975 as a crewmember aboard the ASTP mission.

Smith, Andrew. *Moondust: In Search of the Men Who Fell to Earth*. New York: Fourth Estate, 2005. The author interviewed all the remaining Apollo astronauts, seeking to learn how their lives had changed because of the experience. This book is a remarkable statement of the lives of this elite group of Americans. Some remain household names, such as Neil Armstrong, who has carried his celebrity experience with both dignity and honor. Many are unknown to all except the space community. Some are garrulous and easy to talk to, others are aloof and guarded. Smith found that all were fundamentally changed by the Apollo experience.

Stafford, Thomas P., and Michael Cassutt. *We Have Capture: Tom Stafford and the Space Race*. Washington, DC: Smithsonian Institution Press, 2002. This is a fine book that is sure to benefit all readers interested in America's adventure in space. Tom Stafford is one of America's most significant astronauts, although he is less well known than some of the others. Stafford made four spaceflights—*Gemini 6*, *Gemini 9*, *Apollo 10*, and Apollo-Soyuz Test Project (ASTP)—but he was especially significant for his efforts since the 1970s as the unofficial ambassador to the Soviet Union for space, and for his key roles in defining space policy in the United States.

Steven-Boniecki, Dwight. *Live TV from the Moon*. Burlington, ON: Apogee, 2010. The book covers the earliest known proposals of television coverage on lunar missions and the constant battle internal politics placed upon the inclusion of the TV system on Apollo missions. Closely related subjects such as the slow scan conversion and later color conversion are discussed, and overviews are included for each piloted Apollo mission and the role TV played in covering the flight.

Sullivan, Scott P. *Virtual Apollo: A Pictorial Essay of the Engineering and Construction of the Apollo Command and Service Modules*. Burlington, ON: Apogee, 2003. A collection of exceptionally accurate drawings of Apollo hardware.

Swanson, Glen E., ed. *"Before This Decade Is Out . . .": Reflections on the Apollo Program*. Washington, DC: NASA Special Publication-4223, 1999. A collection of oral histories with some of the key individuals associated with Project Apollo, including George Mueller, Gene Kranz, James Webb, and Wernher von Braun. Rpt. 2002 by University Press of Florida.

Thomas, Andrew R., and Paul N. Thomarios. *The Final Journey of the Saturn V*. Akron, OH: University of Akron Press, August 2011. The Saturn V can be considered one of humankind's greatest achievements. Unfortunately, the demise of the Apollo program left the unused Saturn launch vehicles to rot outside, where they became home to flora and fauna. Hoping not only to resurrect the physical rocket, but also to bring the complete Moon adventure back to life, the Smithsonian Institution and other prominent partners laid out plans to create a total "mission experience" destination at Kennedy Space Center. A key component of the plan was the complete restoration of the Saturn V by Paul Thomarios.

Tribbe, Matthew D. *No Requiem for the Space Age: The Apollo Moon Landings and American Culture*. New York: Oxford University Press, 2014. Offers a portrait of a nation questioning its values and capabilities with Apollo as the center of this debate.

Turnill, Reginald. *The Moonlandings: An Eyewitness Account*. New York: Cambridge University Press, 2002. Longtime BBC aerospace reporter Turnill gives a comprehensive overview of the Apollo program, including its origins in America's post-*Sputnik* panic, the preliminary Mercury and Gemini programs, the drama of the *Apollo 11* landing and the *Apollo 13* near-disaster, as well as the program's demise amid waning public interest, rising costs, and a general sense that the Moon launches had accomplished all they could accomplish.

Webb, James E. *Space Age Management: The Large-Scale Approach*. New York: McGraw-Hill, 1969. Based on a series of lectures, this book by the former NASA administrator tried to apply the concepts of large-scale technological management employed in Apollo to the other problems of society.

Wendt, Guenter, and Russell Still. *The Unbroken Chain*. Burlington, ON: Apogee, 2001. Memoirs are in vogue for the Apollo pioneers. Guenter Wendt was the legendary "pad leader" for all the human space launches from the first Mercury mission in 1961 through the last Apollo flights.

Westwood, Lisa, Beth Laura O'Leary, and Milford Wayne Donaldson. *The Final Mission: Preserving NASA's Apollo Sites*. Gainesville: University Press of Florida, 2017. A discussion of the historic sites of the Apollo program and how they might be preserved.

Wilford, John Noble. *We Reach the Moon: The New York Times Story of Man's Greatest*

Adventure. New York: Bantam, 1969. One of the earliest of the journalistic accounts to appear at the time of *Apollo 11;* a key feature of this general, undistinguished history is a sixty-four-page color insert with photographs of the mission. It was prepared by the science writer of the *New York Times* using his past articles.

Wilhelms, Don E. *To a Rocky Moon: A Geologist's History of Lunar Exploration*. Tucson: University of Arizona Press, 1993. This detailed account of lunar exploration and science strikes a balance between personal memoir and history. As history it provides an exhaustive and contextual account of lunar geology during the 1960s and 1970s, and a less comprehensive detailed but informative account for the rest of the century. As memoir it provides an engaging story of the scientific exploration of the Moon as seen by one of the field's more important behind-the-scenes scientists.

Worden, Al, and French Francis. *Falling to Earth: An Apollo 15 Astronaut's Journey to the Moon*, New York: Smithsonian Books, July 2011. As command module pilot for the *Apollo 15* mission to the Moon in 1971, Al Worden flew on what is widely regarded as the greatest exploration mission that humans have ever attempted.

Young, John W., with James R. Hansen. *Forever Young: A Life of Adventure in Air and Space*. Gainesville: University Press of Florida, 2012. An astronaut's personal experiences in Gemini, Apollo, and beyond.

Zimmerman, Robert. *Genesis: The Story of Apollo 8*. New York: Four Walls Eight Windows, 1998. A detailed account of the December 1968 circumlunar mission to the Moon of Frank Borman, Bill Anders, and Jim Lovell.

Film and Video Works

America in Space: The First 40 Years. 1996. Finley-Holiday Film Corp. 51-minute video general history of space exploration by the United States.

Apollo 11: A Night to Remember. 2009. Acorn Media. Paul Vanezis, director. 118-minute DVD. Using rare archival footage from the BBC, this two-hour documentary compiles the sights, sounds, and electrifying drama of humanity's first footsteps on the Moon. Astronomer Sir Patrick Moore and veteran newsmen cover events as they happened from the launchpad in Cape Kennedy, Mission Control in Houston, and the BBC desk in London.

Apollo Moon Landings: Out of This World. 1996. Finley-Holiday Film Corp. 56-minute video providing a general narrative of the Apollo program.

Apollo 17: Final Footprints on the Moon. 2012. Midnight Pulp Productions. 50-minute DVD and instant download. A tribute to three astronauts and the thousands of men and women behind them during the final days of NASA's Apollo program.

Apollo 13. 1995. Feature film directed by Ron Howard and produced by Brian Grazer. Screenplay by William Broyles, Jr., and Al Reinert. One of the best feature films ever made about the U.S. space program, this work captures the dynamism and drama of the near-disastrous mission without sinking to hagiography or mendacity. Tom Hanks as astronaut Jim Lovell and Ed Harris as mission controller Gene Kranz stand out in a fine ensemble cast. Unlike most Hollywood productions, this work paid close attention to historical detail and captured the reality of the mission.

Apollo 13—Journey to the Moon, Mars, and Back. 2006. NOVA, Noel Buckner and Rob Whittlesey, directors. 270-minute DVD. Tells the gripping, true story of the catastrophic flight of *Apollo 13* and the heroic struggle to bring the astronauts back alive. With firsthand accounts from the pilots, their families, and the people of Mission Control, it documents a thrilling struggle against time and odds and serves as a reminder that, in the words of James Lovell, "We do not realize what we have on Earth until we leave it."

Apollo 13—NASA's Historical Film. 1995. Finley-Holiday Film Corp. 60-minute video history of the mission originally produced by NASA not long after the flight but rereleased in VHS format for educational institutions.

Apollo 13: To the Edge and Back. 1994. WGBH Boston. Written, produced, and directed by Noel Buckner and Rob Whittlesey. 56-minute video history of the mission.

First Man. 2018. Feature film directed by Damien Chazelle. Screenplay by Josh Singer. Based on the book by James R. Hansen, this biopic of Neil A. Armstrong depicts the first Moon landing.

For All Mankind. 1989. 80-minute documentary film produced and directed by Al Reinert. Deals with the Apollo missions, and uses only actual visuals from the missions and the narratives of the astronauts on the missions.

History of Spaceflight: Reaching for the Stars. 1996. Finley-Holiday Film Corp. 60-minute video history of NASA hosted by Alan Shepard.

In the Shadow of the Moon. 2008. VELOCITY / THINKFILM. 110-minute DVD by David Sington. Film vividly communicates the daring and the danger, the pride and the passion, of this era in American history. Between 1968 and 1972, the world watched in awe each time an American spacecraft voyaged to the Moon. Only twelve American men walked upon its surface, and they remain the only human beings to have stood on another world. The film combines archival material from the original NASA film footage, much of it never before seen, with interviews with the surviving astronauts, including Jim Lovell (*Apollo 8* and *13*), Dave Scott (*Apollo 9* and *15*), John Young (*Apollo 10* and *16*), Gene Cernan

(*Apollo 10* and *17*), Mike Collins (*Apollo 11*), Buzz Aldrin (*Apollo 11*), Alan Bean (*Apollo 12*), Edgar Mitchell (*Apollo 14*), Charlie Duke (*Apollo 16*) and Harrison Schmitt (*Apollo 17*). The astronauts emerge as eloquent, witty, emotional, and very human.

Man on the Moon with Walter Cronkite. 2009. 2-DVD set. CBS. Presents the 1969 Moon-landing telecast.

Mission to the Moon. 1986. Signature Productions. Directed by Christine Solinski. Written and produced by Blaine Baggett. 56-minute video on the Apollo program narrated by Martin Sheen.

Moonshot. 1994. TBS Productions. Produced and directed by Kirk Wolfinger. 200-minute dramatization, with archival footage, of the history of the human spaceflight program since the 1950s, hosted by Barry Corbin.

NASA: 50 Years of Space Exploration. 2006. Madacy. 1,026-minute DVD collection. These five DVDs present the most thrilling moments in U.S. space exploration, while also examining the heartbreaking events when tragedy struck the astronauts. Includes Mercury, Gemini, and Apollo, as well as Skylab, ASTP, Space Shuttle, and the space station through NASA's fifty-year history.

One Giant Leap. 1994. Barraclough Carey Productions for Discovery Network. Directed by Steve Riggs. Produced by George Carey. Documentary on Project Apollo.

The Right Stuff. 1983. Feature film directed by Philip Kaufman and produced by Irwin Winkler and Robert Chartoff. Screenplay by Kaufman, based on the book by Tom Wolfe. A cast of relative unknowns at the time depicted the development of aeronautics and astronautics from 1947 through the Mercury program. Scott Glenn as Alan Shepard and Ed Harris as John Glenn captured the essence of being an astronaut. A box-office hit, the film also won four Academy Awards.

To the Moon and Beyond . . . 1994. SunWest Media Group. 56-minute video discusses Apollo program and recent history of space exploration.

When We Left Earth—The NASA Missions. 2007. Discovery Channel. 258-minute DVD. To celebrate fifty years of incredible achievements, the Discovery Channel partnered with NASA to reveal the epic struggles, tragedies, and triumphs of the human spaceflight program. Along with the candid interviews of the people who made it happen, hundreds of hours of never-before-seen film footage from the NASA archives—including sequences on board the actual spacecraft in flight—were carefully restored, edited, and compiled for this collection.

Index

Index